SPECTROSCOPIC
TRICKS

VOLUME 2

From:
APPLIED SPECTROSCOPY
Tricks and Notes
Especially Reorganized and Rearranged for this Edition

SPECTROSCOPIC TRICKS

VOLUME 2

Edited by
LEOPOLD MAY
Department of Chemistry
The Catholic University of America
Washington, D. C.

PLENUM PRESS • NEW YORK • 1971

The material contained in this volume originally appeared in sections of Tricks and Notes in *Applied Spectroscopy* from 1966 through 1969, and is reprinted here by permission of the Society for Applied Spectroscopy.

Library of Congress Catalog Card Number 67-17377

ISBN-13: 978-1-4684-1736-4   e-ISBN-13: 978-1-4684-1734-0
DOI: 10.1007/978-1-4684-1734-0

©1966-1969 Society for Applied Spectroscopy
Softcover reprint of the hardcover 1st edition 1917

©1971 Plenum Press, New York
A Division of Plenum Publishing Corporation
227 West 17th Street, New York, N.Y. 10011

FOREWORD

Spectroscopic Tricks was introduced in 1959 as a special section in the journal *Applied Spectroscopy*. Its purpose was to provide a means for communicating information on new devices, modifications of existing apparatuses, and other items of this nature of interest to the working spectroscopist. That it has proved valuable is indicated by the continuing publication of this section now under the title of *Spectroscopic Techniques*. However, the usefulness of these contributions, scattered through the many issues of the journal, diminishes as time passes since the reader must consult the annual indices of many volumes of the journal to find the contribution that may hold the solution to his problem. The collection of the contributions into a single volume for the years 1959 through 1965 made it easier for the reader to make this search. The success of the first volume has prompted the continuation of these collections.

The contributions in this second volume are selected from the the years 1966 through 1969. They are arranged in the same manner as in the previous volume according to the area of spectroscopy. Those concerned with the same devices are placed together so that the reader can compare them readily. To maintain the advantages inherent in a single collection of articles, the subject index for this volume includes all the entries and page references from the original volume. Both author and journal indices are are also provided, the latter citing the original *Applied Spectroscopy* edition.

The use of the contributions has been approved by the Society for Applied Spectroscopy, whose cooperation in this matter is gratefully acknowledged.

Leopold May

CONTENTS

INFRARED SPECTROSCOPY

MASS SPECTROSCOPY

MISCELLANEOUS

SECTION 1

EMISSION AND ATOMIC ABSORPTION SPECTROSCOPY

A New Briquetting Technique 1.1

Charles K. Matocha

Alcoa Research Laboratories, New Kensington, Pennsylvania

A briquetting technique which overcomes one of the major problems of pellet preparation, that is, seizure and galling of the die assembly, has been developed for the preparation of powder samples. A new product—"Spec-Cap" is utilized in the briquetting technique. The Spec-Cap, shown in Fig. 1, is a 30-mm diameter shell with a $\frac{5}{16}$-in. skirt of 0.0125-in. aluminum and an exterior coating of lacquer. In preparing the pellet, the cap is filled with powder and then centered in a $1\frac{1}{4}$-in. diameter briquetting mold assembly. As the diameter of the Spec-Cap (30 mm) is less than the diameter of the mold (31.8 mm), the metal, upon pressing, forms a slight roll-over or lip on the pressed face. The powder is, therefore, totally contained within the cap and not allowed to extrude between the walls of the die. The height of the cap is reduced in size depending upon the compactability of the powder. The coating of lacquer on the exterior

 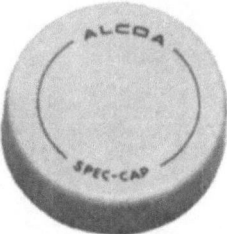

FIG. 1. Spec-Cap.

FIG. 2. Sample of Bauxite prepared with Spec-Cap, 1¼-in. diam.

acts as an antifriction surface and prevents die seizure. Unpainted caps require application of lubricant
to either the cap or die. The thin-gage aluminum allows maximum energy for compacting the sample
rather than compressing the metal. The pellets have
uniform diameter. Spec-Caps of 0.009- and 0.0125-in.
aluminum were tested, and the latter was selected
for use. Although the roll-over or lip is not as large
with the thicker aluminum, the pellets exhibit greater
rigidity. Figure 2 shows a sample of bauxite prepared
with the Spec-Cap. Sample weight for this material
averages 4.9 g.

The briquetting technique was originally developed
for preparation of nonmetallic, powdered samples
for x-ray fluorescence analysis using the Philips PW
1210 x-ray analyzer. The sample loading cups take

1¼-in. diameter sample disks with the sample masked to expose 1⅛-in. diameter. Restrictions were thus imposed on the briquetting technique used for sample preparation. Techniques using an aluminum ring to contain the sample with subsequent pellet formation between open-faced jaws of the press were unsatisfactory, because of the uncontrolled peripheral dimension. Binder-backed techniques used to obtain rigid pellets were considered too time consuming and have the problem of die seizure. Spec-Caps eliminate these objections and produce briquets of high density and extreme rigidity, with protected edges. The latter features are advantages when inserting compacted powder samples in a vacuum x-ray spectrometer having inverted sample optics.

The technique was later applied to preparing wafer compacts of aluminum powder. Preparing a compact of this product in the laboratory is difficult. Again, with the sample contained totally within the cap, these powder compacts offered no preparation problems.

Among the most significant advantages gained through the use of the Spec-Cap is the ability to analyze the aluminum powder compacts using point-to-plane, high-voltage spark excitation. Previously established technique required remelting the sample with a flux and casting a sample disk. This procedure is accurate but requires considerable time. A Spec-Cap briquet can be produced in minutes. Figure 3 shows an aluminum powder compact that has been subjected to a high-voltage, spark discharge in air and in a nitrogen atmosphere, and to a low-voltage

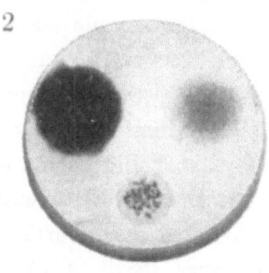

Fig. 3. Aluminum powder prepared with Spec-Cap. (1) High-voltage spark discharge in air. (2) High-voltage spark discharge in nitrogen atmosphere. (3) Low-voltage discharge in air.

discharge in air. A clamp acts as a heat sink and facilitates sample handling. Briquets that can be subjected to a high-voltage spark discharge have. also been prepared of nonmetallic samples mixed with graphite.

Pelleting pressures are not critical and a wide variety of pressures have been tested. The maximum pressure tested was 65 200 psig. Although satisfactory briquets can be produced at considerably lower pressures, a pressure of 40 750 psig has been selected for use. This pressure can be obtained on both motor-driven and manually operated hydraulic presses having a capacity of 25 tons. To eliminate entrapped air and to allow the roll-over to form properly, the pressing cycle is started and pressure is increased to 8000 psig and held for about 10 sec. The pressure is then increased to 40 750 psig in about 10 sec and held there for about 20 additional seconds. The pressure is slowly reduced and the briquet is removed from the mold. Although shortcuts can be taken, this procedure has produced excellent results over a wide variety of both nonmetallic and metallic powders.

The Spec-Cap, therefore, offers a direct, inexpensive method for preparing briquets of a variety of pow-

dered samples. The procedure avoids the major preparation problems—die seizure and galling. Considerable interest has been shown in the development of
this technique, and the Spec-Cap is being made
commercially available through distributors of spectrochemical supplies.

Preparation and Spectrochemical Analysis 1.2
of Lung Tissue for Beryllium*

H. L. Hayes, W. J. Bisson, and W. H. Dennen†

*Cabot Spectrographic Laboratory, Massachusetts Institute of
Technology, Cambridge, Massachusetts*

The usual means of preparing organic tissue for
trace-element studies involves either a wet- or dry-
ashing procedure. Unfortunately, neither method is
entirely satisfactory since contamination may be difficult to control in wet ashing and it yields volumi-
nous products, while dry ashing may allow the escape
of volatile compounds.

One solution to this problem of sample preparation
appears to be dry ashing in a situation whereby both
early-evolved volatiles and the more refractory components are available for analysis. This can be accomplished by spectrographic means if a specially
designed electrode is employed.

Sample preparation starts with the formalin-pre-
served tissue. This is diced, weighed, and placed along
with enough demineralized water ($\sim$15 cm^3/g) to

*This study has been supported by funds from AEC Contract
AT(30-1)-2629.
†Present address: Department of Geology, University of Ken-
tucky, Lexington, Kentucky 40500.

make a slurry in a high-speed tissue homogenizer when run for about 5 min. The slurry is then transferred to a small vacuum jar immersed in a constant 40°C water bath and is pumped down to about 10 mm of mercury.

Dehydration to a flaky or fiberboard-like texture occurs in 2 to 4 h/g of wet tissue with a weight reduction of about 5.5 to 1 (wet to dry). The dehydrated tissue is then reduced to a fine powder in a blender mill (Pitchford Scientific Instrument Corporation) operated for about 5 min.

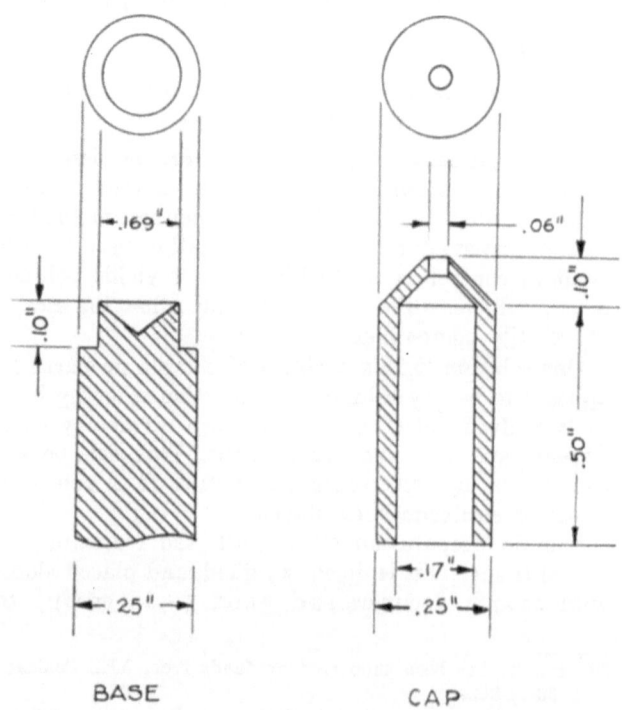

BASE   CAP

Fɪɢ. 1. Electrode and cap for tissue analysis.

The dried lung powder produced by vacuum dehydration cannot be satisfactorily arced in an open-crater electrode because the high electrode temperature causes the sample to be volatilized in a matter of seconds. Consequently, a "boiler type" electrode (Fig. 1) has been designed and used which spreads the evolution of volatile material from the sample over a period of about 30 sec, thus allowing highly volatile material to be introduced slowly and uniformly into the analytical gap. Volatilization of the more refractory residue follows, and arcing is continued until the sample is exhausted.

The electrode consists of two parts, a base and a perforated cap. The design features which are of particular importance to the successful functioning of this electrode are the large electrode mass which reduces the rate of sample heating, the small aperture in the cap which concentrates the early-evolving gases into a jet, the thin walls of the cap which burn away rapidly during the intermediate arcing stage, the small crater in the base which collects the residual bead, and the snug fit of base and cap which eliminates gas leakage.

The sample electrodes described are machined from $\frac{1}{4}$-in.-diam spectrographic grade graphite rods. The counter-electrode is a $\frac{3}{16}$-in.-diam spectrographic grade carbon rod which has been pointed in a pencil sharpener.

Dried lung powder for analysis is placed in an inverted electrode cap by means of a plastic funnel. The amount of sample, usually 60 to 120 mg, is determined by weighing the cap before and after filling. After filling and weighing, a base is firmly inserted in the cap and the sample is ready for arcing. Just prior to striking the arc, the counter-electrode is put in contact with the sample electrode, blocking the exit hole in its top. The current is then turned on and run until a wisp of smoke escapes from the sample electrode. At this time the electrodes are separated to 7 mm with the sample electrode as the anode and the arc operated at 9 A until the sample is exhausted, about 5 min.

A Littröw-type large quartz spectrograph (Adam Hilger Company E.478) is used for all determinations. This instrument has sufficient dispersion so that the analysis line, Be 2348.62 Å, is easily resolved from possible interfering lines, notably Fe 2438.30 Å. The external optical path is arranged to provide an image of the source at the collimating lens.

Spectra are recorded on Kodak Spectrum Analysis —1 plates developed for $4\frac{1}{2}$ min at 19°C in Kodak D-19 developer. Plates are calibrated by the rotating-sector method. Optical densities are determined by use of a Hilger non-recording microphotometer. Spectral line intensity values derived from these measurements are corrected for background and for differing sample weights.

Reference standards are obtained by the addition of BeO to beryllium-free lung powder using a standard dilution procedure and as a by-product of the determination of beryllium in diseased lung by the addition method.[1]

The sensitivity of the procedure described is about 0.04 μg/g of beryllium in wet tissue based upon an initial (wet) sample weighing 0.5 g.

The coefficient of variation has been determined to be ±14% of the value reported.

A test of the accuracy of the method described has been made by comparing the results of spectrographic and chemical analysis of the same powdered sample. G. W. Boylen, of the Industrial Hygiene Laboratory at MIT, made the chemical determination using a

1. L. H. Ahrens, and Taylor, S. R., *Spectrochemical Analysis*, 2nd ed., (Addison–Wesley Publishing Co., Inc., 1961) pp. 454.

wet-ashing preparation and fluorometric analysis. For this referee sample, he reports 1.33 $\mu g/g$ and this group 1.51-$\mu g/g$ Be in the dried tissue.

This procedure seems promising primarily because it eliminates steps used in dry or wet ashing which are susceptible to loss or contamination of the element sought. Of lesser importance, but desirable at times when specificity is required in analyzing human tissue for toxic elements, is the ability of the analyst to take large amounts of tissue and produce a homogeneous sample.

The authors wish to acknowledge the assistance given by Dr. Harriet L. Hardy and by G. W. Boylen.

Dilution Device for Atomic Absorption Analysis 1.3

Leland R. Crawford, Jr., and Thomas Greweling

Department of Agronomy, Cornell University.
Ithaca, New York 14850

One of the advantages of atomic absorption analysis is the very high sensitivity of this technique. However, this sensitivity is not always desirable since time is required to dilute samples too concentrated to be measured. Dilution is also required for solutions of high salt content to prevent burner clogging.

To overcome these difficulties and to facilitate the addition of competitive cations when determining the alkaline earth metals, a stream splitting device has been utilized. This device consists of a T-shaped capillary tube, one arm of which is connected to the plastic tubing of the atomizer of the atomic absorption unit. The second arm is connected to the sample, and the third to a diluent containing strontium or

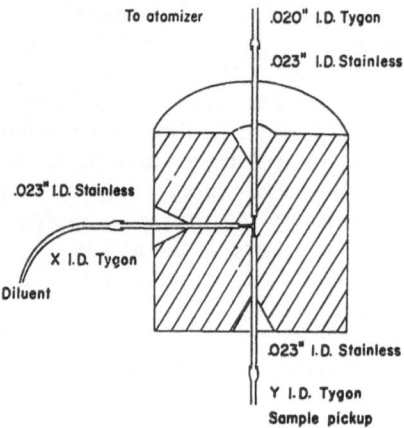

Fig. 1. Dilution device. Dimensions of X and Y tubing regulate dilution; i.e., X = 11 in. long, 0.025-in. i.d., and Y = 2 in. long, 0.015-in. i.d. gives a dilution of about seven to one.

lanthanum. By varying the diameter and length of the tubes to sample and diluent, a considerable range of dilutions may be achieved. When this device was first used there was a considerable surge on the recorder réading when the sample pick-up tube was removed from the sample. This surge necessitated relatively long sample pick-up times to allow a large shoulder to form on the tracing in order to insure accurate readings. More recent tests have shown that this surge may be reduced or eliminated by using a stream splitter having very small capillary tubes and by having some resistance to flow in the diluent tube. Flow resistance may be achieved by having either a long diluent tube or else one of relatively small diameter. An excellent stream splitter having inlet tubes of 0.023-in. i.d. may be obtained from Technicon Controls, Inc., Ardsley, New York 10502. (Stream splitter PT-2 catalog No. 116092, about $32.00.)

Although this device may be utilized manually, it is particularly useful when coupled with automatic sampling equipment such as Technicon's Sampler II and with a strip-chart recorder to record the readings. With this equipment samples may be run at a rate of 120 samples/h. This sampler automatically introduces a water rinse between samples.

The dilution device has been successfully used to analyze soil extracts in 10% sodium acetate solution (Morgan's extracting solution) without prior dilution. The calibration curves remain reproducible from day to day so long as the pick-up tubes do not become clogged. Although at least one manufacturer of atomic absorption equipment (Perkin–Elmer) specifies that the length and inside diameter of the tube to the aspirator are critical, no difficulty has been encountered here when using tubes that vary significantly from the manufacturers recommendations.

The atomizer burner system used here is the spray chamber type which facilitates mixing of sample and diluent. A total consumption type atomizer burner might not produce equally sastifactory results.

Simple Interchangeable Hollow-Cathode 1.4
Lamp for Use in Atomic-Absorption
Spectrometry

G. I. Goodfellow

Atomic Weapons Research Establishment, Aldermaston, Berkshire, England

Hollow-cathode lamps have been used almost exclusively as emission sources in atomic-absorption spectrometry. Although the lamps have been mainly the sealed-off type, two reports by Koirtyohann *et al.*[1,2] refer to an interchangeable cathode lamp constructed from metal with a continuous inert-gas flow. Interchangeable cathode lamps are extremely useful for laboratories engaged in varying types of analysis, particularly of the "one-off" type. This note describes the performance, based on four years of experience, of the flow-through type of lamp shown in Fig. 1, and illustrates the ease with which stable hollow-cathode discharges can be obtained from a variety of cathode materials. The flow-through system avoids the outgassing steps necessary for the preparation of sealed lamps and, therefore, enables cathodes to be quickly interchanged.

The body of the lamp is constructed from glass, with a silica window sealed to the end. Nilo–K rod is used for the metal-to-glass seals because of cheapness, rigidity, and ease of fabrication. None of the dimensions shown in Fig. 1 are critical. Apart from the ease of construction, glass lamps have the advantage over metal lamps that discharge behavior may be readily observed during adjustment of operating conditions.

High-purity argon (99.995%) from a cylinder supply is continuously pumped by a rotary pump (Edwards type 2SC50) through the lamp. The discharge is operated at a pressure of 1–2 mm Hg controlled by a needle valve placed between the cylinder and the lamp. The argon is not normally purified, but, if hydroxyl bands (from water vapor in the cylinder) cause spectral interference, they may be

1. S. R. Koirtyohann and C. Feldman, *Developments in Applied Spectroscopy*, J. E. Forrette and E. Lanterman, Eds. (Plenum Press Inc., New York, 1964), Vol. 3, pp. 180–189.

2. S. R. Koirtyohann and E. E. Pickett, Anal. Chem. **37**, 601–603 (1965).

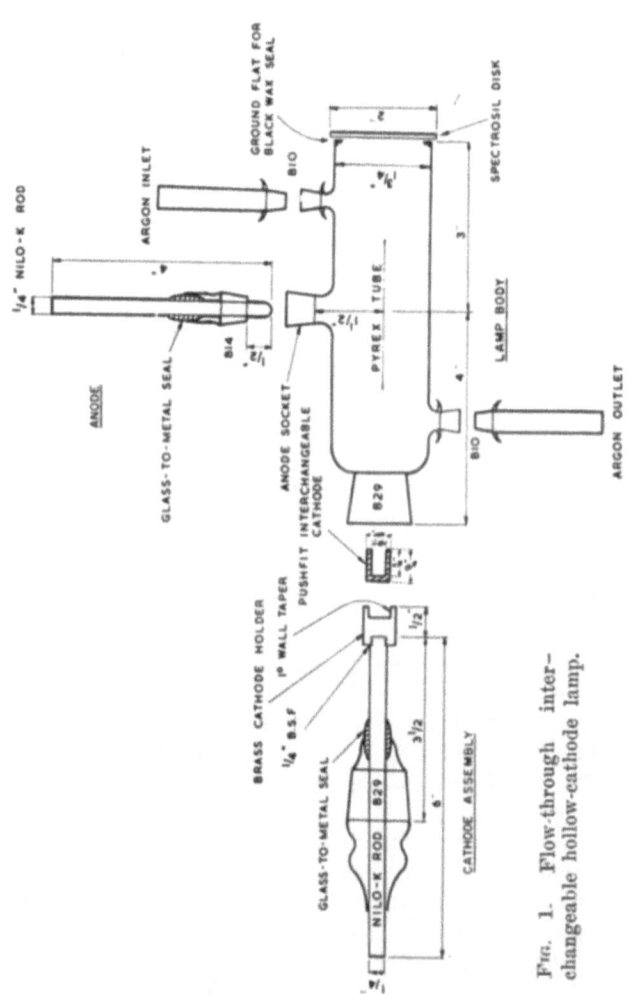

Fig. 1. Flow-through inter-changeable hollow-cathode lamp.

reduced considerably by passing the gas through a liquid-air trap before admission to the lamp. No precise estimates have been made of the running costs of the argon, but indications are that they are very low.

The lamp is operated at the lowest current consistent with good stability and intensity. Currents in the range 20–80 mA are used, the actual value depending upon the cathode material. The flow-through lamp appears to need a higher operating current to obtain the same intensity as the same type of cathode in a sealed lamp; this is possibly due to a cooling effect of the gas passing through the lamp.

Newly prepared cathodes often exhibit surface scintillations for several minutes after starting the discharge, but this unstable period can be shortened by temporarily reducing the pressure below 1.0 mm Hg and then resetting at the chosen operating pressure.

After extended use, a deposit builds up on the walls of the tube adjacent to the cathode, which eventually causes instability in the discharge. When this occurs, the lamp is dismantled and the deposit removed by washing in acid. Since, at pressures below 1.0 mm Hg, the rate of deposition is accelerated, it is desirable to keep above this pressure if frequent cleaning is to be avoided. On the other hand, since intensity progressively decreases with increasing pressure, the operating value should be kept below 2.0 mm Hg if worthwhile cathodic emission is to be obtained.

Apart from some elements such as aluminum and chromium, which form stable oxide layers on the cathode surface, the warmup time to achieve steady emission is usually some 15–30 min. If a change of current or pressure is made, stable emission is reestablished after a few minutes. The time taken for changing cathodes and reinitiating the discharge is less than 5 min.

Five different types of cathodes have been used, enabling sources for atomic-absorption spectrometry to be quickly prepared from a wide range of materials. These are:

1. Solid cathodes fabricated from rods of metal or alloy.

2. Foil-lined cathodes prepared by lining a graphite cathode with foil formed into a cylinder, with a disk of the same foil placed in the base.

3. Plated cathodes prepared by plating a hard coating some 0.005–0.010 in. thick onto a suitable metal cathode.

4. Graphite cathodes loaded with small pieces of metal. The cathode is not filled more than a quarter of the way, and for low-melting-point elements the lamp is tilted slightly to avoid loss of the globule.

5. Graphite cathodes impregnated with metal salts. These are made by heating, plunging into a concentrated salt solution, and drying. Excess crystal growths are removed by smoothing with emery paper before use.

The cathodes of types **1–3** give not only the best short-term stability but also an emission intensity which is substantially constant for a large number of hours. Cathodes of types **4** and **5** may show a considerable decrease in intensity after several hours, due to changes in the inner surfaces of the cathode. In the case of type **4**, this is due to a carbon deposit being formed on the pieces of metal by sputtering from the walls of the supporting cathode. With type **5**, the decrease arises from a gradual loss of salt by vaporization from the surface layers of the graphite. By removing these salt-de-

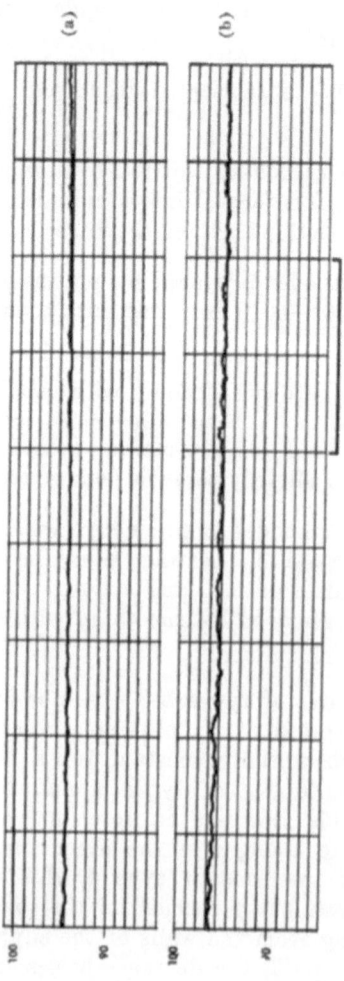

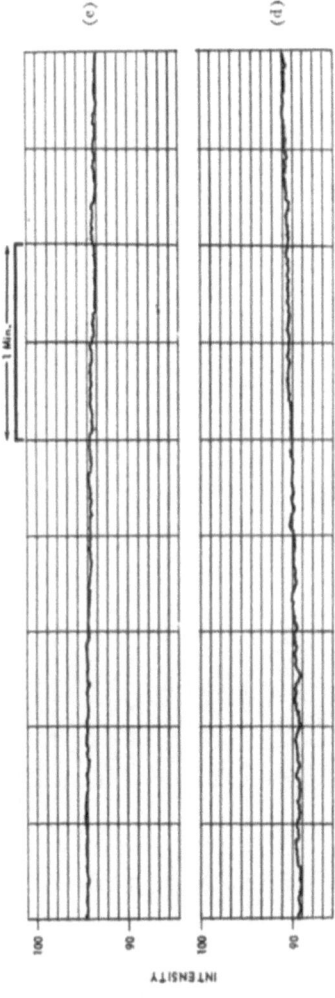

FIG. 2. Chart recordings of emission intensity for sealed and flow-through lamps. (a) V 3184-Å line for a flow-through lamp with a vanadium in graphite cathode, at 45 mA. (b) V 3184-Å line for a Ransley sealed lamp with a vanadium-lined cathode, at 20 mA. (c) Cu 3247-Å line for a flow-through lamp with a brass cathode, at 45 mA. (d) Cu 3247-Å line for a Hilger sealed lamp with a brass cathode, at 45 mA.

Table I. Performance data for different types of cathode.

Cathode type	Element	Description of cathode	Current (mA)	Warmup time (h)	Wavelength (Å)	Stability (%)
1	Cr	Solid Cr cathode	50	$1\frac{1}{4}$	3578	±0.25
	Cu	Brass cathode	45	$\frac{1}{2}$	3247	±0.25
	Cu	(For comparison—Hilger sealed lamp)	45	$\frac{1}{4}$	3247	±0.25
	Fe	Mild steel cathode	45	$\frac{1}{2}$	2483	±0.25
	Mn	"Cuminal" (12% Mn) cathode	37	$\frac{1}{2}$	2794	±0.5
	Zn	Brass cathode	35	$\frac{1}{2}$	2138	±0.25
2	Ag	Ag foil in graphite cathode	45	$\frac{1}{2}$	3280	±0.25
	Ni	Ni foil in brass cathode	60	$\frac{1}{2}$	2320	±0.5
3	Cr	Cr-plated brass cathode	50	$1\frac{1}{4}$	3578	±0.25
4	Al	Al metal in graphite cathode	60	1	3961	±0.25
	Ga	Ga metal in graphite cathode	40	$\frac{1}{2}$	2874	±0.25
	Pb	Pb metal in graphite cathode	40	$\frac{1}{2}$	2833	±0.5
	Sn	Sn metal in graphite cathode	40	$\frac{1}{2}$	2863	±0.25
	V	V metal in graphite cathode	45	$\frac{1}{2}$	3184	±0.25
	V	(For comparison—Ransley sealed lamp)	20	$\frac{1}{4}$	3184	±0.25
5	Ca	Graphite cathode impregnated with $CaCl_2$	74	$\frac{1}{4}$	4226	±0.25
	Ca	(For comparison—Optica sealed lamp)	19	$\frac{1}{4}$	4226	±0.25
	Na	Graphite cathode impregnated with Na_2SO_4	40	$\frac{3}{4}$	5890/6	±0.5

nuded layers with emery paper, the discharge intensity can be restored. In spite of these disadvantages, both of these cathodes are quite suitable for exploratory work.

The performance of a number of different elements using various types of cathodes is shown in Table I. The estimates of stability are based on the range of intensity variation over a 2-min period. This period is considered more than sufficient to establish a base line for absorption measurements taking some 15–30 sec. In common with sealed lamps, there is often an intensity drift over long periods of time, but this is easily overcome by instrumental adjustment. The performance of three commercially available sealed lamps under similar conditions of amplification and excitation is also given in Table I.

In Fig. 2, chart recordings are shown which compare the intensity variations from vanadium (a,b) and copper (c,d) lamps of the sealed and flow-through types. It illustrates that the stability of the latter is at least as good as, if not better than, the sealed type. Since variations in intensity closely follow variations in both current and pressure, it is essential to use a stabilized power unit together with a stable gas flow, in order to obtain the best performance from the lamp.

Acknowledgments. The author wishes to thank R. C. H. MacCormac for helpful advice in preparing this note, J. S. Beaver for technical assistance in developing and testing the lamp, and D. A. Henson for its construction.

1.5   Hollow-Cathode Lamp for Use in Emission and Absorption

R. A. Woodriff

Montana State University, Bozeman, Montana

G. V. Wheeler and W. A. Ryder

Idaho Nuclear Corporation, Idaho Falls, Idaho

Demountable hollow-cathode lamps have frequent application as sample emission sources in high resolution spectroscopy[1,2] and as light emission sources in atomic absorption spectrometry.[3,4] We are interested in both of these uses and also the application of the lamps for the sample absorption sources in atomic absorption spectrometry.

Desirable features of a demountable hollow-cathode lamp for continuous gas flow operation include: (1) the cathode should be open on both ends and the lamp should have windows on both ends so that it may be used in absorption and, in addition, as an aid in optical alignment; (2) the discharge should operate from both ends of the cathode to give increased intensity, and, if used in absorption, increased sample vapor in the path; (3) cathodes, containing samples, should be easily and rapidly exchangeable; (4) cathodes should be simple to manufacture; (5) cathodes should be easily cleaned; and (6) gas flow should be such that sputtered metal is carried away from the windows.

Figure 1 illustrates a hollow-cathode lamp designed to satisfy these requirements. The body of the lamp

1. J. M. McNally, Jr., G. R. Harrison, and E. Rowe, J. Opt Soc. Am. **37**, 93 (1947).

2. T. Lee and L. H. Rogers, Appl. Spectry **15**, 3 (1961).

3. G. Rossi and N. Omenetto, Appl. Spectry. **21**, 329 (1967).

4. G. I. Goodfellow, Appl. Spectry. **21**, 39 (1967).

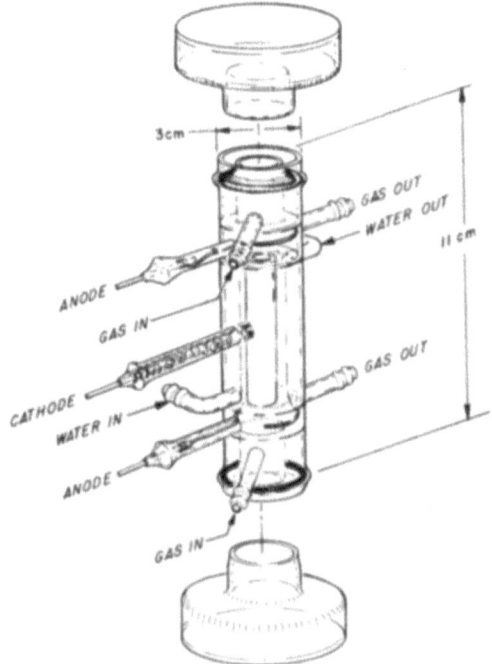

FIG. 1. Demountable hollow-cathode lamp.

is made of borosilicate glass or silica. The windows are
silica, fused or cemented to short pieces of tubing
which make vacuum seals to the lamp body by sliding
into O-rings. Light losses from distortions where the
windows are sealed to the body are reduced by using
windows with a diameter larger than the body of the
lamp. The filler gas enters the annular regions of the
lamp body, flows over the windows, and then toward
the cathode to help keep the windows clear of sputtered
metal. The cathodes are made by cutting appropriate
lengths ($\sim$5 cm) of tubing or by drilling holes in rods
of approximately 1-cm o.d. Note that if there is more
than about 1-mm clearance between the cathode and

the lamp body, a discharge will occur outside the cathode to the electrical lead-in. The cathode lead-in spring holds the cathode in place and provides the electrical contact. Starting cooling water before the discharge is started has prevented any difficulty due to contact of the hot cathode with the cooled walls of the water chamber.

None of the lamp dimensions are critical. Use of a hollow-cathode lamp of this design and size with appropriate power and gas supplies has demonstrated satisfactory emission and promises interesting applications in absorption.

Acknowledgments—This work was done under the sponsorship of the Associated Western Universities at the United States Atomic Energy Commission National Reactor Testing Station. The authors thank H. G. Brinkley for advice on design and for construction of the lamp.

1.6   Versatile Hollow-Cathode Light Source for Spectrochemical Analysis in the Vacuum Ultraviolet*

Giulio Milazzo
Istituto Superiore di Sanità, Rome, Italy

The development of the spectrochemical analysis of nonmetals progressed much slower than that of the metallic elements. This difference could probably be traced to the following reasons: (1) lack of sensitive spectroscopic lines in the visible or near ultraviolet region; (2) poor spectral sensitivity; (3)

*This research was made possible through the support and sponsorship of a contract with the U. S. Department of the Army through its European Research Office.

excitation difficulties; and (4) materials to be analyzed are often poor conductors and mechanically unworkable as electrodes. Some of these difficulties can be overcome by utilizing the vacuum ultraviolet spectral region (VUV), where the nonmetallic elements show their most sensitive spectral lines (i.e., resonance lines). The remaining difficulties can be ameliorated if an exciting source is provided, which allows these elements to be used as electrodes.

This paper discusses the design and general features of the hollow cathode, which was found to be a versatile light source for use in the VUV. The reasons for the choice of this light source and detailed results obtained with it will be published in a later article.

The following requirements must be satisfied by a hollow-cathode light source if it is to be effectively used for spectrochemical analysis:

• It must be capable of operation with every possible cathodic emitting material, either as a massive hollow cathode or as a charge into a hollow cathode of another material.

• It must be simply operated.

• It must be easily changed without destroying the optical alignment.

• It must provide a "gas window" to hinder, as far as possible, the escape of material being tested, which is contained in the cavity as a powder.

• It must be capable of operation at the highest possible current intensity in order to reduce the exposure times.

• It must be cooled.

• It must be able to be operated with flowing or resting gas.

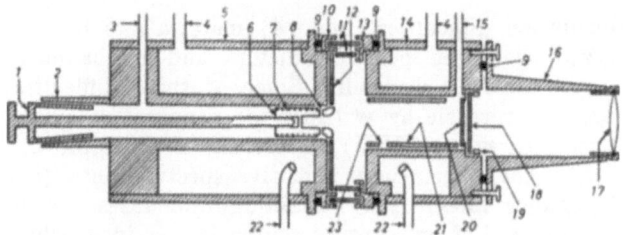

FIG. 1. Hollow cathode source. (1) Standard conical fitting, male, shaft bearing part; (2) standard conical fitting, female part; (3) carrier-gas outlet; (4) cooling-water outlet; (5) cathodic block; (6) hollow cathode; (7) and (8) channels for carrier gas; (9) sealing O ring; (10) connection ring, cathodic side; (11) insulating glass rings; (12) insulating quartz disk; (13) connection ring, anodic side; (14) anodic block; (15) carrier-gas inlet; (16) standard conical connection with the spectrograph; (17) LiF lens (if desired); (18) vacuum-tight-closing LiF window; (19) fastening ring for 18; (20) protecting LiF window; (21) cylindrical fastening for 20; (22) cooling-water inlet; and (23) extracting hole for 21.

• It must be precision machined so that it can be very easily mounted and aligned reproducibly along the optical axis of the spectrograph.

• It must be easily connected with, and separated from, a vacuum spectrograph.

• It must be easily demounted, so that it can be cleaned of cathodic sputtered films.

An improved version of the Newbound and Fish[1] light source was finally used. The apparatus is illustrated in Fig. 1. It consists of a cathode (5) and an anode (14), and can be cooled by water or other coolant material.

1. K. B. Newbound and F. H. Fish, Can. J. Phys. **29**, 317 (1951).

The anode and cathode blocks are connected together by means of two insulating glass rings (11) and held by connecting metal rings (10 and 13). The outer glass ring is tightly fastened in the corresponding grooves by means of Araldite® seals. The inner ring is loosely fitted and can be easily removed. Its purpose is to collect the material sputtered from the cathode during the operation and to avoid its deposition as a metallic conducting film on the outer, tightly fastened, insulating ring. The hollow cathode (6) is fitted tightly in the axial cavity of the anode block, so that it can be efficiently cooled. It too can be easily replaced by removing section 1, which fits vacuum tight into the standard conical joint (2). The hollow cathode contains the exit channels (8) for the carrier gas, located on the front portion of the cavity. This allows a "gas window" to form on the cavity face, which hinders the escape from the cathode cavity of the materials being tested.

The anodic block is axially perforated to permit the observation of the spectrum emitted by the hollow cathode, without need of other optical devices— e.g., reflecting mirror, prisms, etc. The entire source is closed on the front side by two optical windows (18 and 20). Window 18 is held by the fastening ring (19), is permanently mounted, and is vacuum tight. The second window (20) is introduced in the anodic block and held parallel to Window 18 by means of a sliding cylinder (21). The second window serves to collect the material sputtered by the cathode, which continuously reduces the transparency of that window. As soon as the transparency of the window is diminished, it can be removed and replaced by a new, clean window without destroying the vacuum tightness of Window 18. These windows can be chosen of different materials (i.e., glass, silica, lithium fluoride) depending on which spectral region one is interested in investigating. The conical joint (**16**) serves to connect, through an air-free path, the source with the spectrograph—and to act as a support for a condensing lens (**17**), if this is desired.

The carrier gas enters through Inlet **15** and proceeds through the hollow cathode and exits at Outlet **3**. The inlet and outlet are connected with gas-purifying and -circulating equipment, which is discussed below. All of the permanent metal points of the light source are of machined stainless steel. In order to avoid electrical discharge from the anode to the cathode block, which would bypass the hollow cathode, a silica disk (**12**) is placed on the surface of the cathode block.

The gas filling, purification, and circulation device is shown in Fig. 2. The mercury diffusion pump sucks the gas from the cathode side, compresses it, and sends it towards the anode side, from where it follows a path indicated by the arrows in Fig. 2 (i.e., pump, ballast, cold trap, hollow cathode, cold trap,

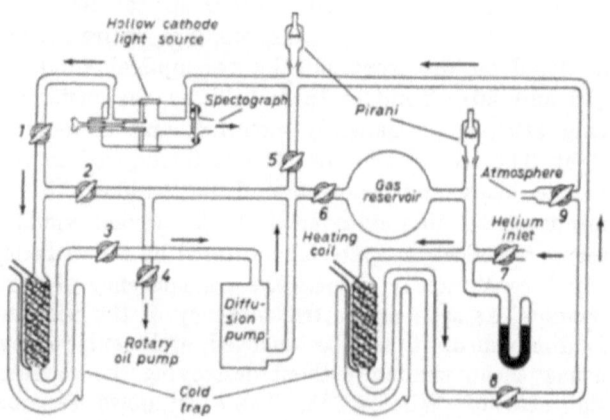

Fig. 2. Schematic diagram of gas circulating and purifying system.

pump). The openings in the gas loop provide for the inlet of the carrier gas and for the connection with the atmosphere. The actual pressure is indicated either by a U manometer or by a Pirani manometer. The hollow cathode can be operated in three ways:

• With resting gas. It will be evacuated from air by connecting to the rotating oil pump, flushed once or several times with the appropriate gas, filled at the desired pressure, and operated without circulation. All stopcocks shown in Fig. 2 are open with the exception of Stopcocks **4**, **7**, and **9**.

• With continuously flowing gas. The carrier gas enters through the inlet and is continuously eliminated by means of the rotary pump. The pressure is regulated by a throttling needle valve inserted between the diffusion and rotating pumps; the diffusion pump is closed off from the system. Stopcocks **1**, **2**, **4**, **7**, and **8** are open—(all others are closed, and the path followed by the gas can be determined from Fig. 2).

• With purified circulating gas. In this case both pumps are in operation. After initial evacuation and flushing, Stopcock **4** is closed to disconnect the system from the backing pump. A gas loop is formed by opening Stopcocks **1**, **3**, **6**, and **8**—(all others remain closed). The carrier gas is continuously purified of mercury vapor and impurities liberated by the light source, by two cold traps filled with activated charcoal and kept at liquid-nitrogen temperatures. The mercury diffusion pump serves as a circulating pump. The traps are provided with electrical heating to allow outgassing when desired, and for regeneration of the activity of the charcoal. Any suitable gas at any desired pressure can be used.

To change the hollow cathode, all stopcocks are closed and the hollow-cathode lamp opened to the atmosphere by Stockcock **9**. After replacing the hol-

low cathode, Stopcock **9** is again closed and Stopcocks **5** and **4** opened, connecting the lamp with the
oil pump. This eliminates the atmospheric gases and
evacuates the system to final pressure. The last traces
of atmospheric gases are eliminated by the diffusion
pump when Stopcock **5** is closed and Stopcocks **1** and
3 are opened. After complete evacuation, the apparatus is ready for use as before.

For the situation where the material to be analyzed
is not conducting or is unworkable in the form of a
massive hollow cathode, another material can be used
as the cathode—such as graphite or aluminum. Both
materials are easily worked, and graphite has a special advantage in that it shows relatively few spectral lines in the VUV. However, because of its poros-
ity, it has several disadvantages also. It requires a
long period to eliminate all of the air trapped in its
pores. Thus, traces of oxygen are unavoidable, and
reaction products of oxygen formed under the electrical discharge (i.e., CO and CO^+ show strong band
spectra in this region. Spectra obtained by this light
source when the hollow cathode is made from graphite will always show these bands. Further, its porous
structure makes it unsuitable for solutions, since the
penetration of the solution into the pores is not reproducible. Thus, the intensity of the spectral lines
obtained for solutions is always nonreproducible.
Aluminum is better suited for sample cathodes, since
it does not present the disadvantages of graphite.

Excellent results have been obtained for iodine
analysis, using the hollow-cathode light source. The
limit of detectability of iodine lines in a matrix of
KCl was about 0.01 μg. The quality of spectra is, in
general, good enough that an atlas of hollow-cathode
spectra of the nonmetallic elements in VUV have
been obtained.[2]

2. J. Junkes, E. W. Salpeter, and G. Milazzo, *Atomic Spectra
   in the Vacuum Ultraviolet from 2250 to 1100 Å. Part I:
   Al, C, Cu, Fe, Ge, Hg, Si, H_2* (Specola Vaticana, Città del
   Vaticano, Rome, Italy, 1965).

Stabilization of Nitrous Oxide/Acetylene Flames

1.7

R. J. Julietti and J. A. E. Wilkinson

*Morganite Research and Development Ltd.,
Battersea Church Rd., London S.W. 11, England*

We have recently found a simple and inexpensive way of stabilizing the flow of nitrous oxide in atomic absorption spectrophotometry. This gas is now commonly used as a support gas for the determination of elements which form refractory oxides, e.g., aluminum, silicon, and titanium. Since the atomic absorption of these elements is critically dependent on the flame conditions, small variations in the composition of the gas mixture lead to erratic readings. Thus when determining percentage amounts of these elements, e.g. in aluminosilicates, it is essential to maintain better control than is needed for minor constituents.

At the high flow rate of nitrous oxide which is normally used, the expansion of the gas in the regulator causes considerable cooling (Joule–Thompson effect). This has an appreciable influence on the performances of the regulator (which was probably designed for use in anaesthetics where the flow rate is much lower). It is possible that the diaphragm loses flexibility at the lower temperature and becomes slow to react to pressure fluctuations.

A desk lamp, with a 100-W bulb, placed a few centimeters above the regulator effectively overcame the chilling, and we now obtain steadier readings.

1.8   Rapid Loading Stallwood Jet

A. W. Fagan and H. M. Klein

Texas Instruments Incorporated, Dallas, Texas

The commercial models of the Stallwood Jet are useful in controlling background, stabilizing arcs, and increasing the sensitivity of certain elements. However, these devices are generally awkward to use since the apparatus must be disassembled to change samples.

To make the Stallwood device practical for routine analysis the basic design of Schribner and Margoshes[1] was modified to permit easy electrode loading. A turn-bolt operated sliding collet was substituted for the fixed electrode holder. The "up" position permits easy loading and unloading and the "down" position automatically positions the electrode in the optical axis. To provide a stable platform to support the sample

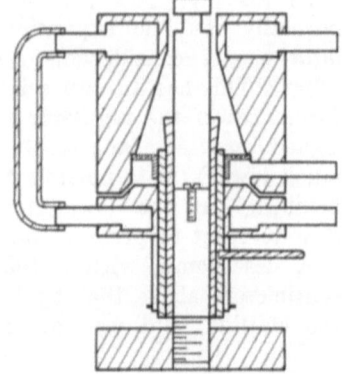

FIG. 1. A cross section of the apparatus.

1. B. F. Scribner and M. Margoshes, N. B. S. Technical Note 272, p. 6, (N) (1965).

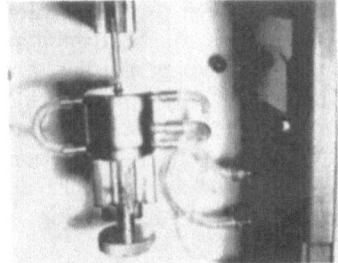

Fig. 2. The apparatus
mounted in an arc stand.

electrode, the apparatus was designed to be screwed
together and can be semipermanently locked into
most commercial arc stands. Figure 1 is a schematic
drawing of the device; Fig. 2 is a photograph of the
apparatus mounted in the arc stand.

Simple Jet for Emission Spectrography   1.9

W. H. Dennen and W. H. Blackburn

*Cabot Spectrographic Laboratory, Department of Geology,
University of Kentucky*

The Stallwood jet[1-3] has found wide applicability
in spectrochemical analysis for the suppression of
selfabsorption, stabilization of the arc, improvement
of signal-to-noise, etc. For the occasional user, however, its cost or complex construction could rule out
this extremely useful accessory.

This laboratory has employed for many years with
complete success a very simple device of different

1. B. J. Stallwood, J. Opt. Soc. Am. **44**, 171 (1954).
2. D. M. Shaw, O. Wickremasinghe, and C. Yip, Spectrochim.
   Acta **13**, 197 (1958).
3. D. M. Shaw, Can. Mineralogist **6**, 467 (1960).

design which provides the same hollow cylindrical gas flow as the Stallwood jet.

The basic principles are that (1) gas flowing through closely spaced orifices expands and coalesces on the low pressure side and (2) gas streams tend to follow the configuration of solid surfaces with which they are in contact. Consequently, gas jets emerging through a ring of small orifices near the

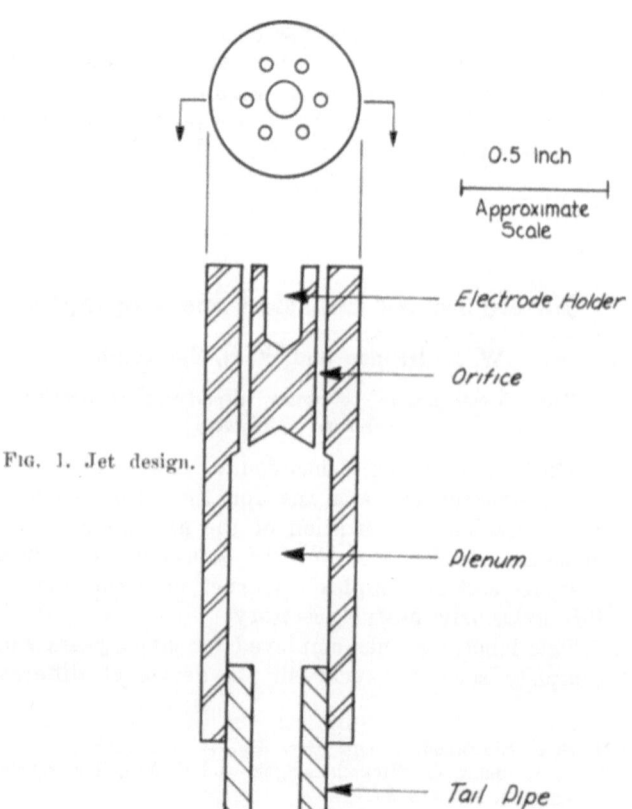

0.5 inch

Approximate Scale

Electrode Holder

Orifice

Fig. 1. Jet design.

Plenum

Tail Pipe

base of an electrode will coalesce and flow up its side providing a hollow cylindrical gas sheath at the analytical gap.

Our design for such a jet device is shown in Fig. 1. No dimensions are given since our experience shows that a wide latitude in dimensions is permissible. Any metal may be employed for construction of the device, and it may be readily machined using only a drill press.

The jet may be employed to deliver air, argon–oxygen, helium, etc., gas flows around the analytical gap for either an open or enclosed arc. Very satisfactory 1-atm gas chambers may be made by adding a glass or silica-glass cylinder as appropriate, closed

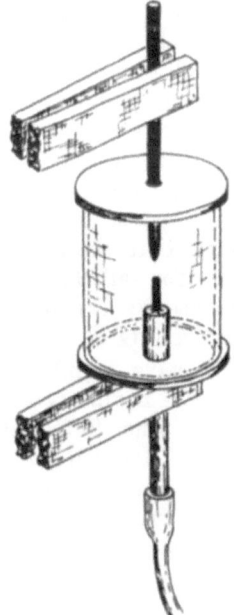

Fig. 2. The jet used with a simple arc enclosure.

by metal covers as shown in the sketch, Fig. 2. The chamber is supported on the lower electrode holder and a slide-fit hole for the counter electrode in the upper cover allows gap adjustments to be made.

1.10   Versatile Gas-Metering System for Controlled Atmospheres

D. L. Nash

Bell Telephone Laboratories, Incorporated, Murray Hill, New Jersey 07971

The use of controlled atmospheres in emission-spectrographic analysis has proved to be of great value in reducing background and in improving sensitivity for some elements. In most cases, the recommended atmosphere is a mixture of two gases such as argon and oxygen. One way of obtaining such mixtures is to purchase a cylinder of gas containing a definite ratio of the two gases desired. This method, however, allows no flexibility in varying the ratio of the gases. Investigations at this laboratory have shown that there is no one gas mixture that will work for all analyses. Best results can only be achieved by varying the gas mixture to obtain optimum results for the particular problem.

We have found a very effective way of providing complete flexibility of gas ratios by using a four-tube Brooks Sho Rate 150 rotameter kit made by the Brooks Rotameter Company, Hatfield, Pa. [Fig. 1(a)]. Each stage is equipped with a needle valve to control the flow. The R-12-15-C metering tubes equipped with a stainless-steel and a Pyrex float cover the range of flows that are encountered in controlled-atmosphere work. Instead of the normal serrated hose

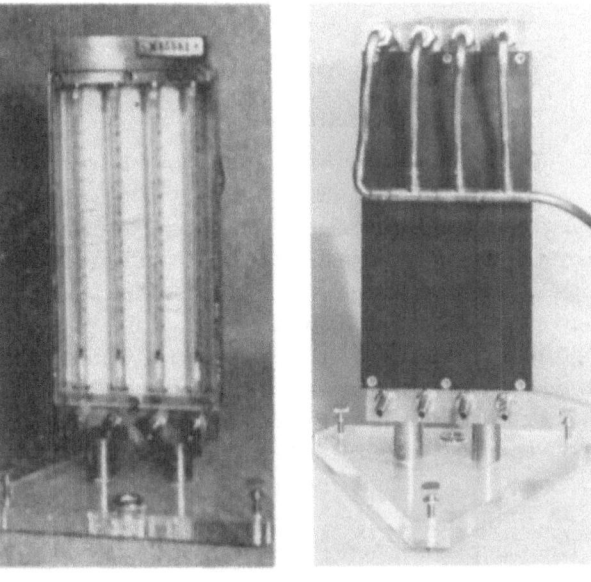

FIG. 1. (a) Gas-metering system showing four flowmeter tubes.
(b) Gas-mixing manifold at rear of metering system.

connections on the output of the meters we used an
adaptor with $\frac{1}{4}$-in. female pipe thread. Using $\frac{1}{4}$-in.
copper tubing and the appropriate copper sweat fit-
tings, a four inlet manifold is made and connected to
the four outputs of the metering tubes by means of
Swagelock fittings as shown in Fig. 1(b). The manifold is connected to the atmosphere chamber with
Tygon tubing.

1.11   Stable Source for Plate Calibration*

Morris Slavin

*Chemistry Department, Brookhaven National Laboratory,
Upton, Long Island, New York 11973*

A recent study of a commercial low-pressure quartz mercury lamp[1] reported by Childs[2] indicated that, with some modification, it might be suitable as a source for emulsion calibration as ordinarily done for the photometry of emission spectra. Childs measured the absolute emission of the Hg 2537 Å line and then determined the intensities of some 55 minor lines, from 5790 to 1942 Å, relative to 2537 Å. All his tests were made with an ac supply of about 300 V at 15 mA.

The lamp consists of a double-bore quartz tube with the electrodes brought out at one end and sealed in a bakelite base. The over-all dimensions are about $\frac{3}{8}$ by 5 in. and the weight is a few ounces. An assortment of shields, for eye protection and for heat insulation, is available, or one can easily be fabricated with any shape of aperture desired.

Childs stated that once thermal equilibrium had been reached, the lamp was very stable. This property of stability combined with an ordinary rotating-sector disk would provide a very effective source for plate calibration. The great attractiveness of

---

*Research performed under the auspices of the U. S. Atomic Energy Commission.

1. These lamps carry the catalog No. SC–I and may be obtained from: Black Light Eastern, Westbury, L. I., N. Y.; Orion Optical Corp., 322 Main Street, Stamford, Conn.; Ultra-Violet Products, Inc., San Gabriel, Calif.
2. C. B. Childs, Appl. Opt. 1, 711–716 (1962).

the rotating sector as a light modulator is that, when run above the critical frequency,[3] its transmittance is simply the percent aperture, which can be read with accuracy on a scale and does not require a previous photometric calibration. However, because of the stroboscopic effect, an ac source cannot be used. The problem thus was to determine the behavior of the lamp when operated on dc.

Experimental. To test stability, the lamp was set up on the optical bench of a 3.4-m grating spectrograph with photomultiplier–recorder readout. The line observed was 5460 Å. A bakelite shield with a slit aperture covered the lamp. Power was supplied by a commercial voltage-regulated dc supply that could be adjusted from zero to 2000 volts. The lamp was operated in two modes, at high and low voltage, with an appropriate current-limiting resistor in series.

For the high-voltage test, the lamp was operated at about 1250 V with a 20-W, 50 000-Ω wire-wound resistor. This passed about 10 mA, which gave good intensity and was well within the safe operating range. Ignition was automatic because of the inductive kick of the wire-wound resistor. The drop across the lamp, after heating up, was 270 V.

Recording of the emission was from a cold start. The lamp reached a stable level in about 6 min; it remained constant during the next 30 min until the test was terminated. The strip chart could be read to an error of about 0.3%, so this was a test not only of the lamp, but, incidentally, of the amplifier as well.

3. J. H. Webb, J. Opt. Soc. Am. **23**, 157–169 (1933).

In the low-voltage mode, the lamp was supplied at 350 V with a variable resistance and milliameter in series. It had to be started with a Tesla coil. Equilibrium was reached in about 5 min and emission remained constant thereafter, except for a slow drift that was also indicated by small changes in current. Even this disappeared after about 20 min running. In the interim, this drift could be compensated by small adjustments of the resistance. Although stability was not as good as with high voltage, constancy of emission could still be maintained to well within 1%, and the smaller supply is an advantage.

One further test was performed—the determination of change in emission with change of current— in order to establish the degree of regulation necessary for a given stability. The current was varied in 1 mA steps over the working range of the lamp and measurement was taken when equilibrium was reached after each change. The emission was found to be nearly proportional to current; it cannot be exactly linear because the relative intensities of the lines change slightly with the current level.

Applications. The intensity of the lamp is high enough to permit short exposures with slow spectrographs and slow emulsions. As a guide, using the minor line at 3022 Å and an SA–1 plate in the second order of a 3-m grating, the exposure to reach a density of 1.0 required 8 sec.

It should be noted that, because of the double-bore construction, the lamp does not emit uniformly in a radial direction. For use as a reproducible source, therefore, the radial orientation must be fixed.

In the application to plate calibration, the simplest procedure is to use a single line, and make a series of graded exposures with variable sector and constant time, which results in an intensity-scale H & D curve. The sector becomes inaccurate at small apertures, but the range can be extended by means of an absorbing filter that, if the series of filtered points overlaps the unfiltered ones, need not be calibrated. For routine work, this procedure may be too tedious; probably it would be more convenient to use it as a primary standard to calibrate a line group, either of the iron-bead arc or of lines of the base element of the samples being analyzed.

Many other uses for a constant dc source should be found in the laboratory. For example, it can be used to determine the reciprocity failure of emulsions, the relative speeds of plates, the extent of scattered light in the densitometer, and the intensity of grating ghosts.

New High-Intensity Spectral Source with a Narrow Line Profile   1.12

Z. van Gelder

*Philips Research Laboratories, N. V. Philips'
Gloeilampenfabrieken, Eindhoven, Netherlands*

In absorption measurement techniques a spectral source is desired that has a resonance spectral line of high intensity. This spectral line must have a profile as narrow as possible without self-absorption. Different solutions of the problem in which a hollow cathode is used are given in the literature.[1-7]

In this note we describe a solution that approaches quite closely the requirements mentioned above. In Fig. 1 a schematic drawing of the electronic circuit of the source tube is given. A positive column dis-

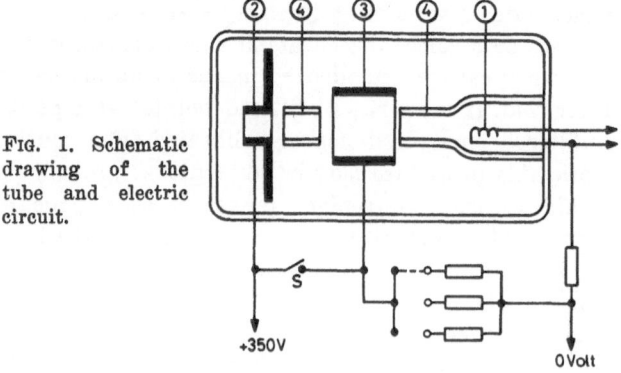

FIG. 1. Schematic drawing of the tube and electric circuit.

charge occurs in a noble gas at a pressure of a few Torr, between (1), the emitting cathode, and (2) a perforated anode disk. Part of the discharge length is surrounded by a cylindrical probe (3). This probe has a negative potential with respect to the plasma potential. Owing to bombardment of the ions extracted from the plasma, atoms are sputtered from the inner surface of the probe. In that part of the column discharge a vapor density is formed of the desired element depending on the composition and structure of the probe surface. Owing to the narrow glass capillary (4), the vapor density decreases very rapidly outside the probe space. The emission of the resonance light in the direction of the axis of the probe is used for absorption measurements.

There are two advantages in this approach. First, the vapor is contained in a well-defined volume and everywhere in the column, where metal vapor is present, electronic excitation takes place, which minimizes the self-absorption. Secondly, owing to the screening effect of the capillaries, the volume from which the light output originates is situated in the axis of the positive column discharge. In this region of the column the density of the exciting electrons has the highest values; whereas the atom density is lowest owing to ionization of the vapor and ambipolar diffusion of these vapor ions towards the probe surface.

These advantages, for the greater part, fulfill the conditions necessary for a spectral line source with a narrow profile and a high intensity. In order to test whether the described construction does indeed lead to a better resonance line profile, the absorption of other sources has been measured and compared with the new tube. This has been done with the help of a copper vapor absorption cell. This cell consists of a small furnace containing a constant vapor density of copper in an argon atmosphere at a pressure of 5 Torr. Such a medium has a narrow absorption profile, which is nearly the Doppler profile at the temperature of the furnace ($\sim 500°C$).

The various tubes which were investigated were: (a) a normal commercially available hollow-cathode tube, (b) a tube based on the principle of the so called "high-intensity tube" of Walsh[6] with a hollow cathode to create the vapor in combination with a positive column, and (c) the new tube. The absorption as measured is defined by $\log I_0/I_1$ called absorbance, where I_1 is the resonance light intensity (at a wavelength of 3247 Å) transmitted through the cell with vapor, and I_0 is the resonance light output of the source without vapor in the cell.

In Fig. 2 the absorbance is given as a function of the output I_0. This source output I_0 is varied by changing the sputtering rate, i.e., changing either the hollow-cathode current or the probe current. The positive column currents in the cases (b) and (c) are kept constant at 200 mA and 150 mA, respectively. [The curves (a), (b), and (c) refer to the tubes mentioned above.]

From Fig. 2 we can conclude that the new tube has an emission profile nearly independent of the

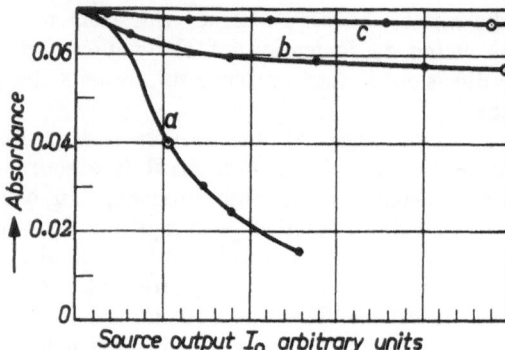

Fig. 2. Absorbance as a function of resonance light output. The optical system with small diaphragms and the absorbing medium (a small furnace containing Cu vapor and argon at a pressure of 5 Torr) are the same for all tubes. The circles give the operating value of the hollow cathode and the other tube constructions with a sputter current of 20 mA.

probe current and thus independent of the produced vapor density. It is nearly the Doppler profile, which can be concluded from the fact that at low sputtering currents the self-absorption in the tubes can be neglected and the major influence on the line profile is Doppler broadening. At very low vapor densities, that is, at very low output of the resonance light, the three tubes indeed give the same value for the absorbance.

Owing to this property still another important advantage of this new tube can be mentioned. It is well known that fluctuations of the light output of the source can be eliminated by using the double-beam method. However, as the results are interpreted by using a calibration curve, the change in absorbance influences the numerical values obtained. With the new tube the change in absorbance is so small that these errors are eliminated.

As is shown in Fig. 1, only one dc source and one ac source is needed to operate the tube. With switch S on for a short time the tube is ignited between cathode and probe. When S is opened again, the discharge starts burning between anode and cathode. As the burning voltage across the positive column is much smaller than the burning voltage between the anode and probe, this probe gets a negative potential with respect to the plasma and sputtering is started by the ions extracted from the plasma.

1. H. Schüler, Z. Physik **35**, 323 (1926).
2. H. Schüler, Spectrochim. Acta **5**, 322 (1952).
3. B. I. Russel and A. Walsh, Spectrochim. Acta **15**, 885 (1959).
4. A. D. White, J. Appl. Phys. **30**, 711 (1959).
5. W. G. Jones and A. Walsh, Spectrochim. Acta **16**, 249 (1960).
6. J. V. Sullivan and A. Walsh, Spectrochim. Acta **21**, 721 (1965).
7. G. J. Goodfellow, Appl. Spectry. **21**, 39 (1967).

Spurious Radiation in Littrow-Mounted Grating Monochromators 1.13

A. Watanabe

Defence Research Telecommunications Establishment, Ottawa, Canada

In this note we discuss two unusual types of "scattered" radiation that can occur in Littrow-mounted grating monochromators. Here, "scattered" radiation refers to diffracted radiation that reaches the exit slit by paths other than the one for which a monochromator was designed. Such spurious radiation usually will be out of focus at the exit slit and may not be noticeable in many applications. How-

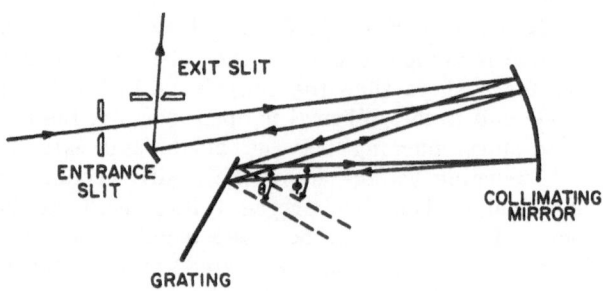

FIG. 1. Path of spurious doubly diffracted radiation in a Littrow-mounted grating monochromator.

ever, its presence can lead to serious error in the measurement of the absorption coefficient of a band, since the intensity at the exit slit for a particular grating angle will be the sum of the desired intensity at the correct wavelength and the spurious intensity at another wavelength.

One of these types of spurious radiation can occur in a single-pass Littrow monochromator. This effect was first encountered by the author in measuring the absorption coefficient of the pressure-induced infrared fundamental band of gaseous hydrogen with a Perkin–Elmer 98G monochromator equipped with a 300 lines/mm grating. The presence of the spurious radiation led to reduced absorption coefficients over most of the band and to rather odd line shapes for the absorption lines. Stamm and Salzmann[1] first observed this spurious radiation when they converted a Perkin–Elmer model 12 prism spectrometer for use as a grating spectrometer. This effect has also been mentioned by Alpert.[2] Nevertheless, it is not universally known, and it seems worthwhile to discuss the origin of this stray light.

1. R. F. Stamm and C. F. Salzmann, Jr., J. Opt. Soc. Am. **43**, 126 (1953).
2. N. L. Alpert, Appl. Opt. **1**, 437 (1962).

The path of the doubly diffracted radiation is indicated in Fig. 1 for a typical Littrow monochromator. For singly diffracted radiation the angle of diffraction θ is nearly equal to the angle of incidence, and the grating equation for radiation of wavelength λ can be written as

$$n\lambda = 2d \sin \theta,$$

where d is the grating space. Some radiation at shorter wavelengths, which is diffracted at an angle $\varphi < \theta$, falls on the collimating mirror nearly normally and is reflected back to the grating.[3] Here, it is rediffracted at an angle equal to θ, and reaches the exit slit. Thus, the equation for the doubly diffracted radiation of wavelength λ_1 is given approximately by

$$n\lambda_1 = 2d \sin (\theta - \Delta)$$

where $2\Delta = \theta - \varphi$. In the Perkin–Elmer model 98G monochromator Δ has a range of values of around $3.5°$ to $5°$.

Double diffraction can occur for radiation that is originally incident on the grating in the region enclosed by surfaces normal to the collimating mirror at its edges. This region on the grating in the Perkin–Elmer instrument is indicated in Fig. 2 by the light hatching. On the second pass, the radiation, which is almost in focus at the grating, is diffracted by the region shown by the dark hatching. The doubly diffracted radiation is out of focus on reaching the exit slit.

The doubly diffracted radiation can be eliminated by means of a premonochromator with a sufficiently narrow passband (less than $\lambda - \lambda_1$). If the spectral

3. Stamm and Salzmann observed the double diffraction for radiation at longer wavelengths, because their grating was mounted so that the zero order lay on the left of the collimator as viewed from the grating, rather than on the right, as is usually the case.

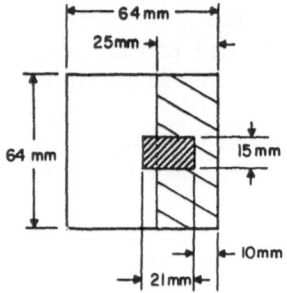

FIG. 2. Regions of the grating utilized by the doubly diffracted radiation in a Perkin–Elmer 98G monochromator in undergoing its initial diffraction (light hatching) and its final diffraction (dark hatching).

region to be examined, in a single scan, is large, the passband must be made to track the setting of the main grating monochromator. A simpler method would be to mask off the area of the grating where the spurious radiation undergoes its initial diffraction, or alternately, the area where it undergoes its final diffraction. Alpert[2] suggested a mask over the former area, which, in the Perkin–Elmer instrument, results in a decrease in intensity of nearly 40%, as well as reducing the resolving power of the grating considerably. On the other hand, a mask over the latter area results in a loss in intensity of less than 10%; this method was adopted in the author's instrument with satisfactory results.

Spurious radiation, resulting from a rather different effect, can reach the exit slit in a double-pass Littrow monochromator of the Walsh type,[4] where a pair of corner mirrors is used to send the diffracted radiation through the spectrometer for a second pass. We have found that, under certain conditions, light that travels through the corner-mirror system in the direction opposite to the designed direction can reach the exit slit, as shown in Fig. 3. For the desired radiation, the first and second pass diffraction occurs in the same order of the grating. For the spurious

4. A. Walsh, J. Opt. Soc. Am. **42**, 94 (1952).

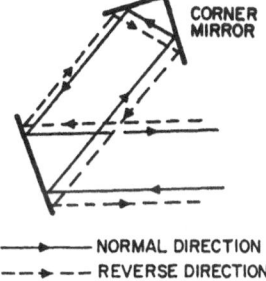

FIG. 3. Path taken by the normal and spurious radiation through the corner mirrors of a Walsh-type Littrow monochromator.

——▶—— NORMAL DIRECTION
— —▶— — REVERSE DIRECTION

radiation, however, the geometry is such that the first and second pass diffraction must occur in two different orders. Thus, the problem exists only when this spurious radiation, at an angle of incidence slightly different from the proper angle, can be diffracted in another order and reach the exit slit. This condition can be satisfied for relatively coarse gratings used in a higher order to diffract short wavelength light. In our particular instrument, the spurious radiation was observed with a 150 lines/mm grating for wavelengths less than 6000 Å, and was also observed in this region with coarser gratings, but not with finer gratings.

This type of spurious radiation could also be eliminated by the use of a tracking premonochromator. However, the difference in the angles of diffraction for the normal and the spurious images was about one degree or less in our instrument, so that this method would be impractical in many applications. A more satisfactory solution would be to avoid the use of coarsely ruled gratings that make it possible for such spurious radiation to reach the exit slit.

Spurious effects, such as the one described above, should be tested for in all instruments. One method is to record the output at the exit slit with a monochromatic source illuminating the entrance slit. Users

of double-beam instruments may find it more convenient to insert in the sample beam a sample or
filter with a strong, isolated absorption line and to
check the zero transmission line. As mentioned previously, the doubly diffracted light is out of focus at
the exit slit. The use of a polychromatic source, as
in the latter method, gives the total stray-light intensity at the exit slit for a particular grating setting and therefore will give a more accurate measure
of the stray-light intensity than with a monochromatic source. A sharp cutoff short-wavelength pass
filter can be used effectively in place of the narrow-
line width absorber or filter, since double diffraction
occurs, in the conventional monochromator, for wavelengths shorter than the wavelength of the singly
diffracted radiation.

1.14 Image-Rotating Device for a Spectrograph Illumination System

C. J. Cremers*

Georgia Institute of Technology, Atlanta, Georgia

E. R. F. Winter

Purdue University, Lafayette, Indiana

Often, in the course of optical investigations, particularly with spectrographs, it becomes desirable
to orient the image of a fixed light source in some
particular direction with respect to the spectrograph

---

*Presently with the University of Kentucky, Lexington, Kentucky.

slit.[1] This situation presents no problem if the light path has a section where the light beam is parallel. Here one may use a Dove prism for the image rotation. However, if the light beam has no section with parallel light or if the optical investigations are being carried out in the ultraviolet or infrared regions of the spectrum where the cost of achromatic optics becomes prohibitive, then one must use front-surface mirrors.

Such a system must contain three reflecting surfaces in order to avoid dislocating the optical axis of the system. The extended normals from the mirror centers must all intersect at a point and these normals must lie in one plane. This plane will also contain the optical axis of the system. A view of this plane is shown in Fig. 1. The solid lines here repre-

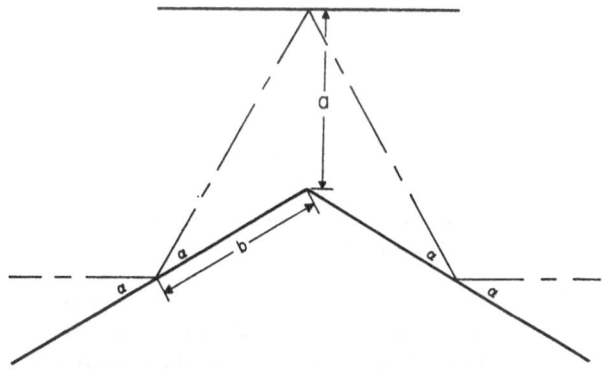

FIG. 1. View of plane containing mirrors.

1. For example, E. R. F. Winter and C. J. Cremers, "Temperature Distribution in a Low Mass Flux Argon Plasma Jet," (USAF, Aerospace Research Laboratories, Wright–Patterson Air Force Base, Ohio.) ARL 62-388 (July 1962).

sent edge views of the mirror surfaces; and the dashed line, the optical axis of the system. For instance, if one wanted to put a horizontal image of a vertically oriented object onto a slit, then the above plane would be at an angle of 45° from the vertical. In all cases, then, the plane will just bisect the angle between the source axis and the image axis. The mirrors shown in Fig. 2 will rotate an image through 90° with respect to its object.

Positioning requirements on the mirrors are:

1. The first mirror should be inclined at such an angle from the optical axis that the component of its area normal to the beam is larger than and encompasses the cross-sectional area of the beam.

2. The second mirror must be parallel to the axis of the undisturbed optical system and centered about the plane of symmetry of the first and last mirrors. It must be spaced from the other mirrors so that the center of the beam where it meets the surface corresponds to the center of the surface. It may easily be shown that this separation a is given by

$$a = b \sin \alpha / (1 - 2 \sin^2 \alpha),$$

where b is the half-length of the oblique mirror and α is the angle of incidence (Fig. 1).

3. The third mirror has the same size requirement as the first. It must be mounted symmetrically to the first with respect to a plane perpendicular to the undisturbed optical axis that passes through the center of the second mirror.

A particular system that is relatively compact is shown in Fig. 3. This may be used for any light beam having a diameter of up to 50 mm. The mirrors are

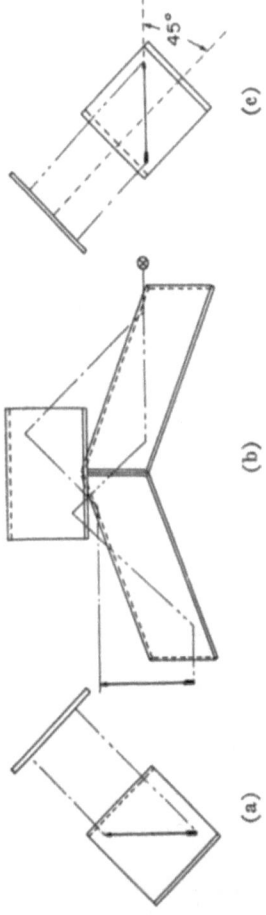

Fig. 2. Schematic front and side views of mirror system for 90° image rotation. (a) Left end, (b) front, (c) right end.

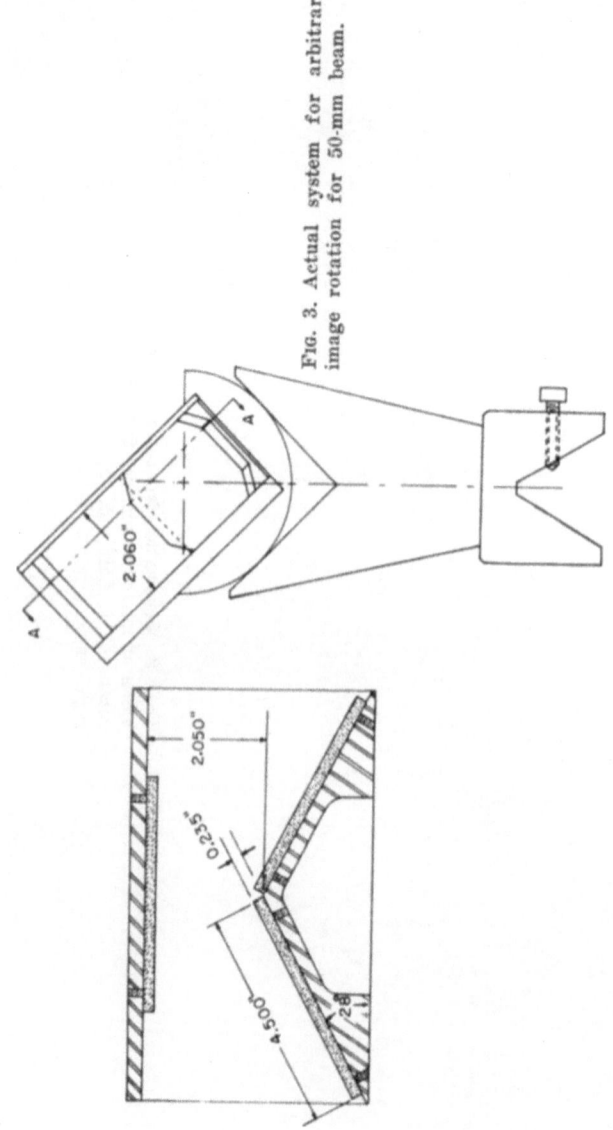

FIG. 3. Actual system for arbitrary image rotation for 50-mm beam.

held at three points by spring clips (not shown) acting against fine-adjusting screws. The device is mounted on a cylindrical pad that may rotate in a V block. This permits adjustment of the desired rotation angle. Not shown are screws that lock the pad to the V block after fine adjustment is made. The mirror assembly is mounted on a carriage which fits a standard 60° optical bench.

Acknowledgment. The authors gratefully acknowledge the assistance of R. A. Gilbert, who carried out the construction of the afore-mentioned device and who also worked out many of the details of its design.

Mask for Selecting Orders for Use with the Jarrell–Ash Order Sorter*   1.15

A. Mykytiuk and S. S. Berman

Division of Applied Chemistry, National Research Council, Ottawa 7, Ontario, Canada

An inherent disadvantage in working at relatively high grating angles in optical emission spectroscopy in order to obtain increased dispersion and higher line-to-background ratios is the presence of multiple orders. If a stigmatic spectrograph (e.g., Jarrell–Ash 3.4-m Ebert) is used, the various orders can be offset or ''sorted'' by a device such as the Jarrell–Ash ''order sorter.''[1] The ''order sorter'' is essentially a low-dispersion spectroscope which projects a vertical spectrum of the source on the slit of the spectrograph. Light of wavelength λ falls on a specific portion of

*NRCC No. 10103.

1. R. F. Jarrell, J. Opt. Soc. Am. **45**, 259 (1955).

the slit while light of another wavelength, for example 2λ, falls on a different portion. If the spectrograph is stigmatic, the two wavelengths produce images at different levels on the photographic plate corresponding to their different positions on the slit; the length of each image being determined by the width of the "order sorter" slit. In this manner, the spectrum is "sorted" and overlapping orders avoided. However, the multiple orders are still in the focal plane to be recorded by the detector.

Owing to the limited wavelength response of detectors, they can often be used also as filters in order to restrict the number of orders detected. For example, using an Ebert 3.4-m spectrograph with a 600 grooves/mm grating set at an angle of 27.25°, Kodak S.A. No. 1 emulsion will record four orders of eight present over a 20-in. focal plane (Table I).

Usually, only one of these wavelength ranges is required (e.g., 5th order, 2875–3325 Å). The presence of the other orders serves only to complicate the presentation and severely limit the amount of useful data which can be presented on the photographic emulsion. The four offset orders shown above occupy 7 mm of plate width resulting in a maximum of 11 possible exposures over a 4-in. plate width.

Table I. Orders and wavelength ranges—27.25°.

Order	Wavelength ranges, Å	Detected by S.A. No. 1
1	14 380–16 620	No
2	7 190– 8 310	No
3	4 790– 5 540	No
4	3 595– 4 155	Yes
5	2 875– 3 325	Yes
6	2 395– 2 770	Yes
7	2 055– 2 375	Yes
8	1 800– 2 080	No

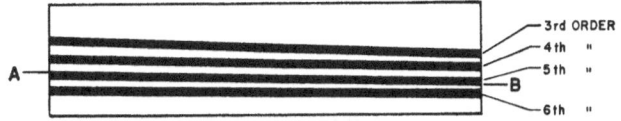

FIG. 1. Diagram of "sorted" orders in the focal plane.

Because it was considered desirable for many studies to present much more than 11 exposures on a single plate, a mask was constructed to eliminate all but the wanted order. This is achieved simply by taking further advantage of the stigmatic aspect of the Ebert mount.

Figure 1 is a representation of the offset orders in the focal plane. It is obvious that a mask could be constructed in front of the photographic plate to allow only the radiation from a particular order to fall upon the emulsion. This would require a mask 20 in. long of rather critical dimensions. However, if we recall that each wavelength range represents a different portion of the spectrograph slit and that the Ebert mount has a 1 to 1 magnification, it is obvious that a mask of width AB placed at the appropriate location at the slit will allow only the desired wavelength range to enter the spectrograph. This is not only easier to produce than a focal-plane mask but should also result in less scattered light in the instrument.

A mask was fashioned (Fig. 2) which could replace the standard fishtail and Hartman-diaphram slide at the spectrograph slit. The vertical height of the aperture is AB (3.05 mm for the 5th order described above). The exact horizontal position is found by trial and error and is indexed by the arbitrary scale on the mask face. Using an "order sorter" slit of 1 mm it is now possible to record 40 exposures, each 1 mm high with a 1-mm spacing on a 4-in. plate.

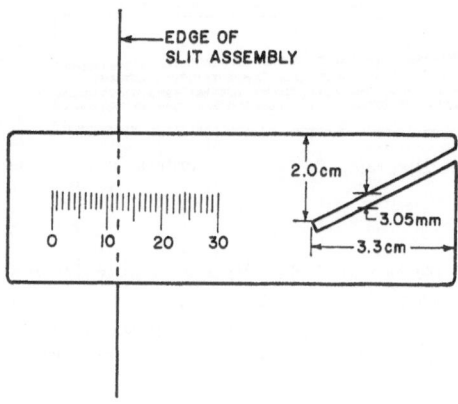

Fig. 2. Mask for isolation of 5th order (2875–3325 Å).

Because of the nonlinear dispersion of the "order sorter" each order requires a different aperture. Also, a mask may be fashioned to incorporate two or more adjacent orders.

1.16   A Device for Making Time Studies

W. H. Dennen

Cabot Spectrographic Laboratory, Department of Geology, University of Kentucky, Lexington, Kentucky 40506

Time studies are used in dc arc emission spectrography to establish certain arcing parameters through observations of the selective volatilization process. The time of entrance of an element into the arc gap, its rate of volatilization, and its time of exhaustion are important facts to be determined when analytical procedures are being set up. Further, time-sampling can provide information as to changing temperatures

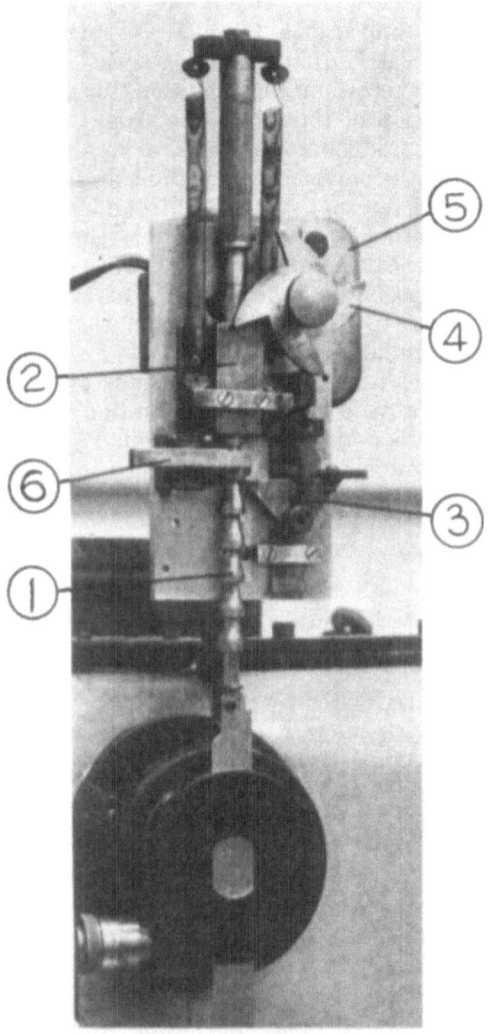

Fig. 1. The device.

in the arc gap and lead to a means of making matrix corrections.

In most applications, the plate is moved at appropriate intervals during the arcing period with successive spectra recording the conditions for each sample. Mechanization of this procedure should improve the data, but reproducible continuous or step-wise cassette or plate-mask movements are difficult to achieve in practice. An alternative approach is to leave the plate fixed and provide for a vertical displacement of the spectrum by causing the entering light to pass through successive portions of the slit. (It is assumed that the slit jaws are perfectly parallel and that the slit is uniformly illuminated.)

Since only tiny vertical movements are needed at the slit to provide the desired separation of successive narrow spectra, the device shown in Fig. 1 and described below employs the difference in escapement travel and hole spacing in a mask to attain closely spaced spectra. As constructed, the escapement teeth have a spacing of 12 mm and the mask carries a series of 0.8-mm holes on 13-mm centers. As the mask is moved, each successive aperture is thus displaced 1 mm from the preceding (aperture separation minus movement per step) and the individual 0.8-mm high spectra are separated by 0.2 mm of clear plate.

The escapement rack (1) was machined on a lathe giving it a circular cross-section which is easily carried through a hole in a bearing block (2). The rack is spring loaded to assure its proper seating against the driving pawl (3). The entire pawl drive is moved vertically in a machined guideway by the timing cam (4). This cam is driven by a 1-rpm motor (5) and different cams may be used to trip the escapement at predetermined submultiples of one minute, for example 10, 15, 20, 30, or 60 sec. Both the driving pawl (3) and the stop pawl (6) may be swung away from the escapement rack to allow its adjustment or return it to the starting position.

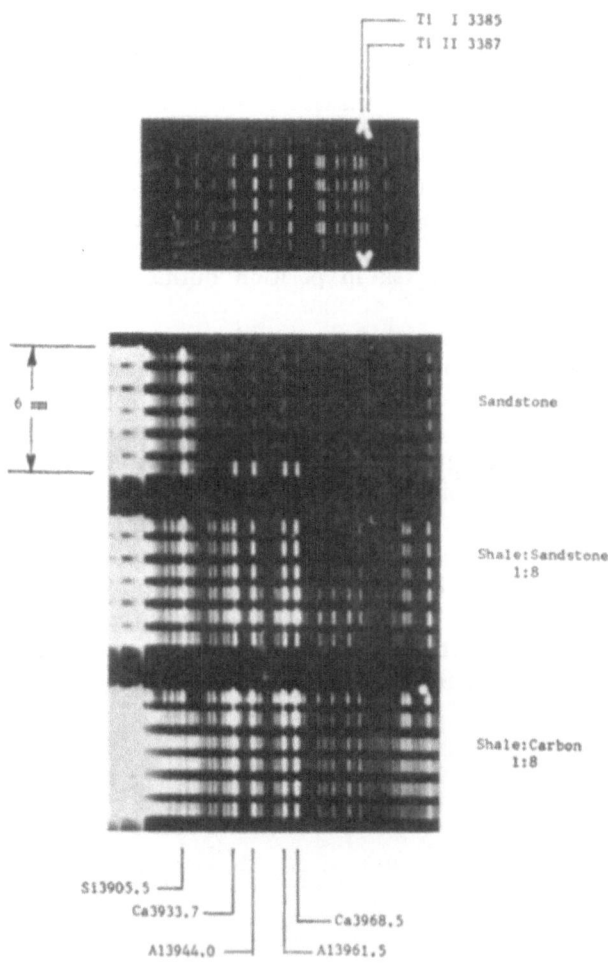

FIG 2 Examples of spectra.

In operation, clockwise rotation of the timing cam forces the pawl drive down its guideway placing a driving spring mounted behind the mounting block under tension. The driving pawl, which is held against the rack by a light spring, drops under the rack a moment before it is released by the timing cam. When the pawl drive is released by the timing cam, the driving spring contracts forcing the pawl drive and escapement rack rapidly upward. At the top of its travel, the spring loaded stop pawl (6) slips into place under a tooth of the escapement rack and holds it and the attached mask in position during the ensuing time interval.

The mask is made from sheet brass and is connected to the escapement rack by a projecting horizontal pin which allows for easy attachment and also slit focusing movement. The mask is carried through the slit cover in a slotted guide machined for this purpose. Fig. 1 shows the entire assembly in place. The entire device, except the detachable mask, is permanently attached to the spectrograph housing, ready for immediate use.

Examples of spectra taken for time study purposes are shown in Fig. 2. Of particular interest is the changing thermal environment of the run as shown by the changing intensity ratio of Ti II 3387 to Ti I 3385.

Slit-Adjustment Mechanism for a 1.5-m   1.17
Bausch & Lomb Spectrometer

J. H. J. Coetzee,* H. J. Jacobs, and J. Morris[†]

South African Iron and Steel Industrial Corporation, Pretoria, South Africa

The direct-reading head of the 1.5-m Bausch & Lomb spectrometer was designed for a fixed analytical program. The setting up of such a program is a time-consuming and highly skilled task. Since it is essential that the program on our instrument be

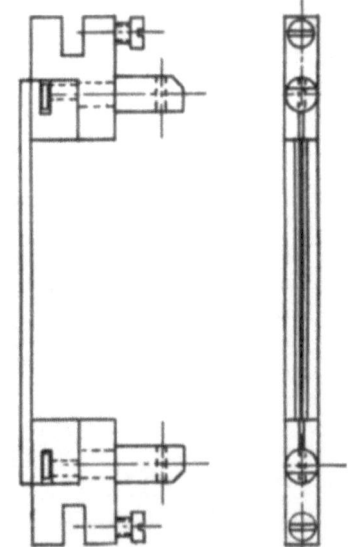

FIG. 1. Original slit.

*Presently at Dept. of Chem., Univ. Pretoria, Pretoria, South Africa.
†Presently at the Natl. Bldg. Res. Inst., S. A. Council Sci. Ind. Res., Pretoria, South Africa.

changed frequently, the original design was unsatisfactory from our point of view.

As supplied by the manufacturer, the slits (Fig. 1)
are clamped to carrier rails (A in Fig. 2) by means
of two screws each, during the rough setting up. The
first coarse positioning of the secondary slits is done
by using an exposed film with the appropriate line positions marked on it, as a template. Final adjustment
of the slits is done photoelectrically by adjusting the
cams incorporated in the slit mount, until the slit
and line are coincident and parallel. This method of
adjustment has a number of drawbacks:

• If the clamping screws are not sufficiently tight,
the whole slit tends to shift or twist when the cams
are rotated.

• If the clamping screws are too tight, the cams
cannot be rotated.

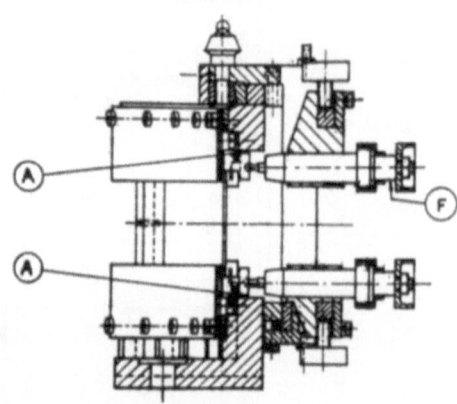

FIG. 2. Section through modified slit frame along the optical
path. Note: all parts of slit frame and slit-adjusting mechanism are to be blackened by chemical process or painted
black except where not required.

• Since there is no provision for locking the screws in position, they tend to loosen after repeated adjustments have been made.

• The upper cams may be adjusted while the direct-reading head is covered and the high-tension feed to the photomultipliers is on, thus giving a continuous indication of the relative positions of line and slit. This cannot be done with the lower cams, since the tool used for the adjustments obstructs the light path, so that it has to be removed before the line can be reprofiled. Whenever the tool is to be removed from, or attached to, the cam, the high tension must be switched off, and since the adjustment of either top or bottom is always accompanied by a slight shift in the other, it becomes a very tedious process indeed.

• In addition, there is the danger that the loose tool and screwdriver, which have to be used for the adjustments, may slip and cause considerable damage to slits and mirrors.

In view of the above difficulties, it was decided to modify the direct-reading head. Ideally, the method of adjustment should be such that the controls are mounted outside the head, but profiling the spectrum entails moving the plate on which are mounted the secondary slit frame and all the postslit optics and photomultipliers. It was therefore decided to design a device which would fit into the direct-reading head, so that it could be operated under the cover of a black cloth while the high-tension supply to the photomultipliers is switched on.

The first modification required was the reconstruction of the secondary slits (Fig. 3), eliminating the cams, and providing spring clamps to hold the slit in position on the rails.

Two rails have been attached to the slit frame (B, C, D in Fig. 4), one on top and the second to the

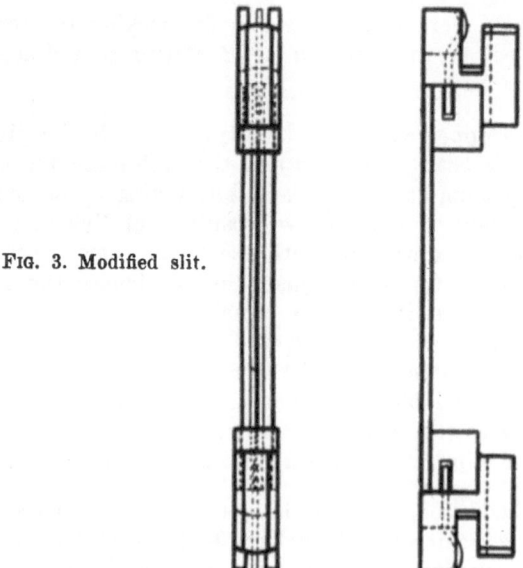

Fig. 3. Modified slit.

lower front edge of the slit frame, as viewed from
the phototube side. A brass carriage (E, Fig. 4) was
constructed to travel on these rails, with a provision
for clamping it in any chosen position. Two steel bars
mounted on dovetail slides on the carriage, each controlled by micrometer screws with differential threads
giving a lateral movement of 100 μ/rot, provide the
adjustment for the top and bottom of the slit.

Mounted on the steel bars are smaller carriages
(Fig. 2), which can be locked in position anywhere
along the bars. Quick-threaded screws (6.6-mm pitch)
(F in Figs. 2 and 5) connect these carriages to the
slit by mating with slots (G, Fig. 4) in the slit
mounts, when these are screwed in, in order that slit
movement may be controlled by the micrometers.

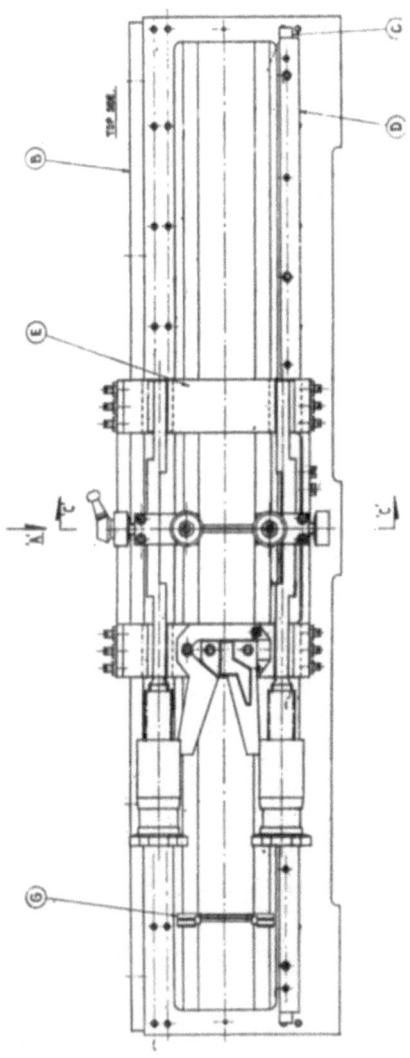

Fig. 4. Modified slit frame as seen from photomultiplier side, showing movable carriage.

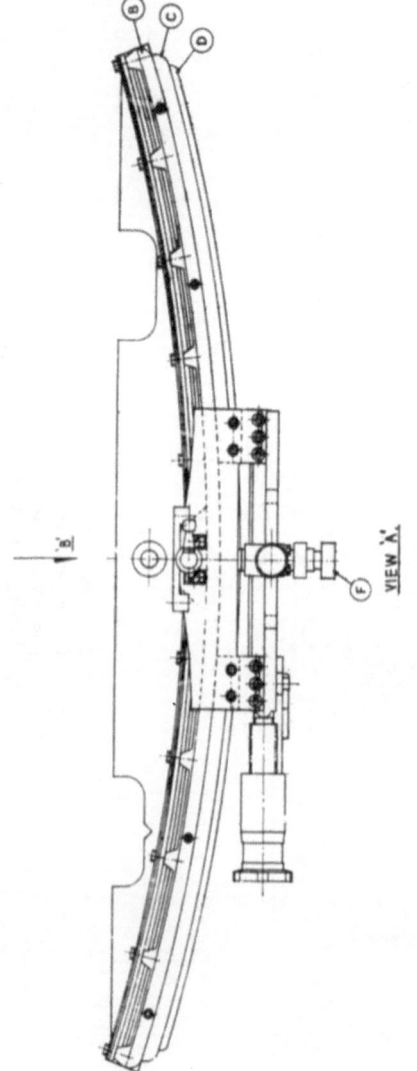

Fig. 5. Modified slit frame as seen from above.

With the main carriage in position and the secondary carriages connected to the slit, the latter may be moved or tilted without the controlling mechanism obstructing the light path.

After positioning the slits approximately, by means of a film template, one slit may be adjusted until it is parallel with the spectral line, and the corresponding profile position noted. Thereafter, the other slits may be checked in turn for parallelism with their spectral lines, and shifted until the profile position of each corresponds with that of the original reference slit.

Since the adjustment of the slit position can be made under a black cloth, while its position is monitored by photoelectric means, it is only necessary to switch off the high-tension feed to the photomultipliers when connecting the adjustment mechanism to the slit or when withdrawing the pins from the slots. In order to avoid moving the slit, it is necessary to back off the micrometers two full turns, thereby ensuring that the pins are completely free in the slots, before withdrawing them.

This modification to the direct-reading head of a 1.5-m B & L spectrometer has turned a strictly routine analytical instrument into a versatile research tool, on which new spectral lines may be selected within a matter of minutes. It may be of interest to note that this comparatively simple modification can be applied with advantage to the ARL Quantovacs, greatly reducing the down time when program changes are made.

Acknowledgment. The authors thank the South African Iron and Steel Industrial Corporation (Ltd.) for permission to publish this paper.

1.18   Attachment for a Rotating Sector

T. A. Read

*Associated Lead Manufacturers, Ltd., Middlesex,
Great Britain*

Because of the marked curvature of spectral lines
obtained with the Hilger medium quartz spectrograph, rotating-sector plate calibration involves repeated manipulation of the microphotometer adjustments to align individual steps with the microphotometer slit.

In a recent program involving the use of a close-
ratio sector, visual discrimination between adjacent
steps proved impossible, and the necessary manipulation became extremely tedious.

The attachment shown in Fig. 1 superimposes a
series of clearly defined lines on the calibration plate.

FIG. 1. Attachment shown mounted in front of the rotating
sector.

Although it was made to cope with the problem quoted, the resulting ease of reading is such that a similar attachment is being made for use with the more usual 2 : 1 ratio sector.

The grooved capstans are eccentrically mounted for precise fore-and-aft adjustment in the 1-mm gap between sector and slit. Height adjustment, which is made visually with the sector rotating, is by simple sliding sleeves.

The threads are of grey nylon monofilament of 0.005 in. diam (1.3 lb fishing line).

There are obvious possibilities of image degradation arising from scattered or fluorescent radiation from the nylon or from the well-known adjacency effects. No such effects could be detected in Kodak B-10 plates after dish development.

Acknowledgment. The author is indebted to Associated Lead Manufacturers Ltd. for permission to publish this note.

Modified and More Versatile Direct-Reading Head for a 3.4-m Ebert Spectrograph   *1.19*

J. Morris* and J. M. C. Van Staden

South Africa Iron and Steel Industrial Corporation, Pretoria, South Africa

A 3.4-m Ebert spectrograph, with its supplementary direct-reading head, was bought for this laboratory with certain requirements in mind:

*Presently at Nat. Bldg. Res. Inst., S. A. Council for Sci. Ind. Res., Pretoria, South Africa.

• It should be a versatile photographic instrument, with high resolution and an optical system that would permit the variation of wavelength ranges and spectral orders, and the interchange of gratings.

• The instrument should provide a direct-reading head as an alternative to the photographic camera such that the changeover from photographic to direct-reading operation should be a simple and rapid procedure.

• It should be possible to alter the direct-reading program at will and with a minimum of trouble.

Photographically, the spectrograph has come up to expectations, and the excellent quality of its spectra makes it eminently suitable for the analysis of ferrous materials. However, the direct-reading head has been disappointing on account of the difficulty of setting up and altering the program. A brief description of the direct-reading head, as supplied by the manufacturer, helps to explain why this is so (Fig. 1).

The secondary slits (1) are mounted on two fixed rails (2), along which they should slide. Because of

FIG. 1. The original direct-reading head.

uneven machining of the rails, the slits bind at some points and move too freely at others. At either end of each slit mount, a hole has been drilled parallel to the optical path, and into this hole a pin may be screwed to connect the slit to the carriage (3) of the adjusting mechanism. The carriage, in turn, is carried on two rods (4), to which it may be clamped at any chosen point, depending on which slit is to be adjusted. The rods are controlled from outside the direct-reading head by means of a plain, uncalibrated knob coupled to a screw (5), which provides a movement of the slit of 0.8 mm/revolution.

A change in direction of movement entails taking up two–three turns of backlash, and the mechanism is not such that the point at which the slit begins to move can be felt. A further complication is that the connecting pin (6) does not always mate exactly with the hole in the slit mount, the result being that the whole system is placed under considerable stress. The force exerted by the adjusting mechanism on the slit tends to rotate the slit about its longitudinal axis, thereby causing it to assume a false position, from which it springs back as soon as the force is removed.

Once the light has passed through the secondary slit, it falls onto a concave cylindrical mirror (7), which focuses it onto the cathode of the photomultiplier tube. As the cylindrical mirrors can be rotated (8) and tilted (9), but not shifted, the number of spectrum lines of any chosen program that will fall through the appropriately positioned secondary slits onto the cylindrical mirrors is severely limited. Those that do not are deflected onto the cylindrical mirrors by means of small plane mirrors (10) mounted directly behind the secondary slits. Choice of suitable reflection angles for the plane and concave mirrors leads to a confusing crisscrossing of light paths, and the increased pathlength and extra reflection reduce the measured light intensity.

All these factors conspired to render the direct-reading head unsuitable for our needs, and it was decided to redesign the head. As far as possible, the existing framework was retained while the head was modified with the following considerations in mind:

• Lateral adjustment of the secondary slit positions should be done by means of calibrated controls. The top and bottom of each slit should be individually controllable, so that absolute parallelism of the slit and the spectral line may be ensured. It should, however, also be possible to move the slit in its entirety, without upsetting this parallelism of slit and line.

• The accuracy and fineness of the adjustment should permit the slit to be positioned reproducibly to within ±2.5 μ of the desired position.

• Disengagement of the mechanism, after the slit has been placed in position, should be possible without fear of the slit position's being disturbed.

• The concave mirrors should be moveable in a lateral sense, so that any spectral lines which may be offset by up to 10 mm relative to the appropriate photomultiplier may be directed onto the photocathode by using only the concave mirrors.

The design and construction of the modifications required to achieve the aims set out above were

FIG. 2. The modified slit-adjustment mechanism.

undertaken by the staff of the instrument workshop of the Research Department.

The slits were modified by using springloaded clips in place of the screwclamps to maintain their position on the two rails. This change means that unevenness in the thickness of the rails does not affect the firmness of the slit mounting as much as was previously the case. The mechanism for moving the slits has been extensively modified (Figs. 2 and 3). The light rods have been replaced by a much heavier square-section rod located on dovetail mounts at either end (Fig. 3). This rod has been curved to the radius of the focal plane, so that the pin (D–D), which now engages in a slit in the end of the slit mount and is carried by a block sliding on the square-section rod, maintains its position in the line of the slit irrespective of the position on the focal curve. As the movement of the slit by means of this mechanism is never more than 5 mm, the rod and the focal curve are essentially concentric at all times.

By using screw threads of 1.0- and 0.9-mm pitch (A–A), which, by difference, give a movement of the slit of 100 μ/rev of the control knob, the top and bottom of the slit can be moved until the slit is parallel to the spectrum line. This is easily done by turning the grating of the spectrograph until a line from a small mercury vapor lamp falls on the appropriate slit. Profiling the first and third steps of the Hartmann diaphragm gives a measure of the skewness of the secondary slit, and the necessary correction can be applied.

At the control end, the rods are mounted on a block which is moved on a dovetail slide by a third control knob (B–B). This knob is calibrated directly in tenths of a micron (10^{-4} mm) and, once the slit has been set parallel to the spectrum line, this control may be used to move the entire slit back and forth without upsetting this parallelism.

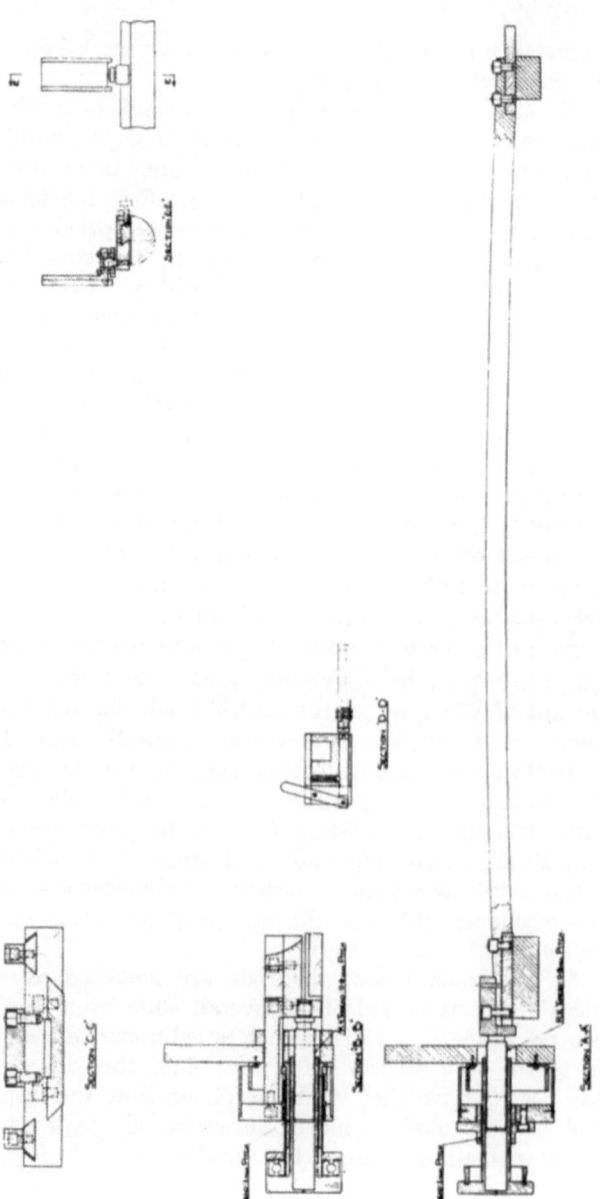

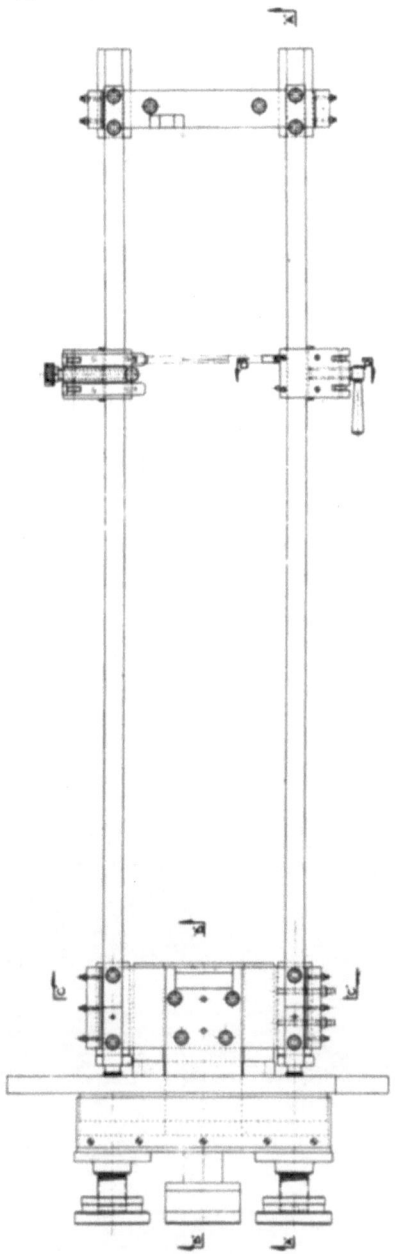

FIG. 3. Diagram of modification.

Because there is still a certain amount of backlash in the system, adjustments are always made in
the same direction. Once the slit is in position, the
main control knob is backed off two and a half turns
to free the pin in the slot of the slit mount, before
withdrawing it. If this is done, the slit is not disturbed.

The bar to which the concave mirrors are attached
has been provided with a rail, along which the mirrors may be moved (E–E, Fig. 3, and Fig. 4). The
mirrors are now mounted on ball joints, which allow
them to be swiveled in any direction. These ball joints
may be mounted on their base blocks in one of two
positions. This permits the staggering of the mirrors
and allows them to be set up much closer together
than would otherwise have been the case. This modification has halved the minimum distance between
lines that may be used without requiring plane de-
flecting mirrors. There is, obviously, a lower limit

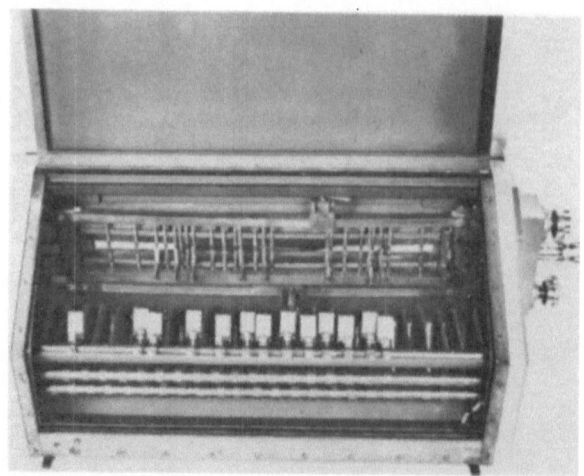

FIG. 4. Complete redesigned direct-reading head.

to the wavelength difference that can be dealt with in this way, and the presence of a third line close to the other two still requires the use of a plane deflection mirror. The fact that photomultiplier tubes cannot be shifted, or even rotated, restricts the versatility of the system considerably. In Fig. 4, the assembled direct-reading head is shown.

A short description of the procedure for setting up a new program may be instructive to those unfamiliar with this spectrograph.

An iron spectrum on a strip of 25-mm film provides a template for the first approximate setting up of the secondary slits. Because the grating can be rotated, one can now use the zero-order reflected beam to set up the postslit optics. A mercury line is included in the program and is the first to be installed. The secondary slit is set up parallel to the spectrum line, and the exact positions of the grating and the entrance slits are noted.

Now, the grating is rotated until the mercury line impinges on another exit slit, and this is adjusted until it is parallel to the spectrum line. The grating is returned to its original position, the profile position of the mercury line is noted, and by using a spark source and a sample containing the element in question, the second exit slit is moved to the same profile position as that noted for the mercury line. This procedure is repeated for each of the other lines in the program.

The modification of the direct-reading head has provided us with a versatile research tool in the place of a routine analytical instrument with a fixed program. We are indebted to our workshop personnel for the excellence of their work, and to Iscor for permission to publish this paper.

1.20   # Test Jig for an Integrating
                          # Direct-Reading Spectrograph

H. G. Yuster and R. J. Hemmer

*U. S. Atomic Energy Commission, New Brunswick Laboratory
New Brunswick, New Jersey 08903*

A Bausch & Lomb dual grating spectrograph was
converted to a direct-reading instrument with the use
of the readout system employed in the Baird As-
sociates Spectromet, model HA-1. The readout system
consisted of four clocks, necessary power supplies,
programming, bias amplifier controls for the clocks,
sequential relays, and charging capacitors.* Since the
spectrograph and the measuring system were not
integral parts, it was necessary to increase the length
of the coaxial signal leads from the multiplier phototube signal system to the readout system, and re-
arrange the electronics in the readout console. With
these changes made in the geometry of the electronics,
it was desirable to test the readout system, prior to
ultimate use of the direct-reading instrument in
spectrochemical analysis. This test would also aid
in the correct setting of the amplifier bias controls,
and indicate malfunction of clocks through a comparison of the four clock readings.

A test jig was designed that enables insertion of
voltages at the ends of the coaxial leads into the
capacitors of the element channels and the capacitor
of the internal standard channel. Since this type of
readout system corrects for phototube dark current
with a polarity reversing relay, it was necessary to
synchronize voltage insertions on the capacitors with
the opening of the shutter on the spectrograph.

---

* Spectromet Instruction Manual, model HA-1, Baird-Atomic,
  Inc. Cambridge, Mass.

A diagram of the test jig is shown in Fig. 1. When the shutter is open, relay RY-1 places the voltages on the capacitors and is actuated by a 1-sec, 12-V pulse received from the programmer. The charging voltages are removed from the capacitors during the 1-sec period of dark current correction when polarity reversing of the capacitors takes place and the spectrograph shutter is closed.

V_1 and V_2 are batteries supplying charging voltage to the voltage dividers R_{is} and R_x. V_1 is the source of charging voltage for the internal standard and V_2 is the source of charging voltage for all of the element or X channels. Their respective charging voltage will be designated as V_{is} and V_x. The voltage dividers were 10-turn heliopots provided with vernier dials each containing 1000 divisions. For the insertions of voltages ranging from 45 V down to 0.45 V, the voltages could be set and read with very good accuracy from the indicated dial readings. Voltage

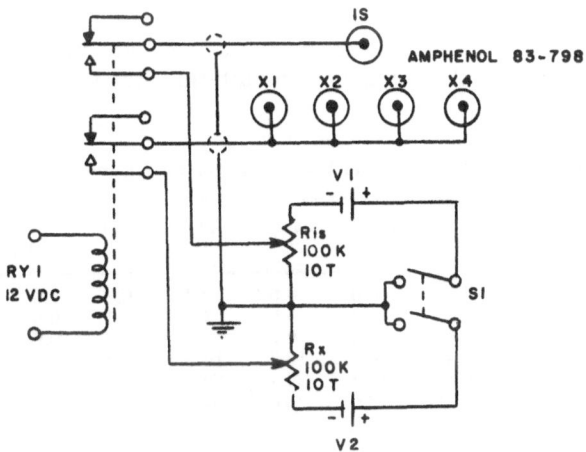

Fig. 1. Wiring diagram of test jig.

Table I. Clock readings for V_{is}=45 V and variable voltages of V_x.

V_{is}	V_x	Clock 1	Clock 2	Clock 3	Clock 4	V_x/V_{is}
45	45	0.0	0.0	0.0	0.0	1.0
45	36	4.1	4.1	4.0	4.0	0.8
45	27	9.4	9.3	9.3	9.3	0.6
45	18	16.8	16.7	16.7	16.8	0.4
45	9	29.5	29.5	29.5	29.5	0.2
45	0.45	42.2	42.2	42.2	42.2	0.1
45	0.18	59.0	58.7	58.9	59.0	0.04
45	0.09	71.9	71.9	71.8	71.8	0.02
45	0.045	84.2	83.8	84.2	84.2	0.01

values below 1 V were established with a Leeds and Northup bridge-type instrument standardized with a standard cell.

Table I shows the results obtained with the internal standard voltage set for 45 V and the resulting clock readings corresponding to various voltages on the X channels. In this table it can be seen that there is very good agreement in the clock readings, which indicates that the clocks were operating satisfactorily and the bias setting sensitivity controls of the amplifiers were good. To eliminate the possibility that the voltage on the internal standard might

Table II. Clock readings for V_{is}=1.0 V and variable voltages of V_x.

V_{is}	V_x	Clock 1	Clock 2	Clock 3	Clock 4	V_x/V_{is}
1.0	1.0	0.0	0.0	0.0	0.0	1.0
1.0	0.8	4.1	4.0	4.1	4.1	0.8
1.0	0.6	9.4	9.2	9.3	9.4	0.6
1.0	0.4	16.8	16.7	16.8	16.8	0.4
1.0	0.2	29.5	29.4	29.5	29.5	0.2
1.0	0.1	42.6	42.8	42.8	42.8	0.1
1.0	0.04	58.7	58.8	58.7	59.2	0.04
1.0	0.02	70.0	69.8	70.0	70.0	0.02

be too high, a second set of runs were made with lower voltages in order to check the system. The data are tabulated in Table II. A 3-V battery was used for V_s to enable more exact settings with the voltage divider R_x.

The data in Table II compare quite favorably with those shown in Table I. It appears evident that a voltage of approximately 50 mV or lower may give erroneous readings. It is questionable whether this is due to lead leakage, capacitor leakage, shorting relay contact resistance, or a combination of the three phenomena.

In the test runs, the total time of the exposure cycle was 30 sec. Therefore, the batteries were in the circuit 15 sec and out of the circuit 15 sec, when normally the dark current undergoes correction. Sufficient time was programmed for resetting of the clocks and for the final readout or measuring cycle of the clocks. In this type of system at the start of the measuring period, a resistance R is placed across the internal standard capacitor C and simultaneously the X capacitor, amplifier, and the clock are placed in series with the RC circuit. The voltage on C will decay to a point where it is equal to V_x and the clocks will continue to run for a time t until this point is reached. Mathematically this process may be denoted by the following equation:

$$t = RC \ \ln \ V_x/V_{is} \qquad (1)$$
$$\text{or clock reading} = K \ \log \ V_x/V_{is}.$$

The above equation shows that if the clock reading is plotted on a linear scale and the value V_x/V_{is} on a log scale, a straight line graph should result. These results are graphically shown in Fig. 2 from the tabulated results listed in Table I. Examination of Eq (1) shows that when $V_x/V_{is} = 1$, the clocks will read zero. When V_{is} is greater than V_x, the clocks will run for a finite period of time, and when V_{is} is

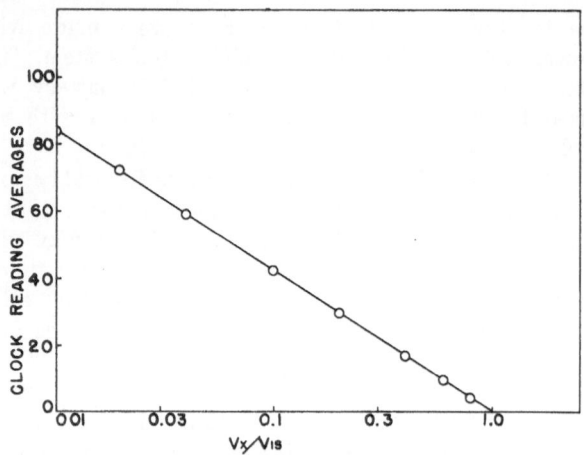

Fig. 2. Graphical presentation of test jig data.

less than V_x, it is negative and the clocks will not start.

This type of test jig is not only applicable to a clock type of readout system, but could be used in a counting tube type. In addition, it would be possible to build the test jig into a direct reader with appropriate switching when an instrument check is desired.

1.21   Baird Clock Revisited

W. R. Kennedy and Robert Norman Smith

American Cast Iron Pipe Company,
Birmingham, Alabama 35202

Baird Atomic Direct Readers,[1] having clock-indicating outputs, frequently present early difficulty in

1. Baird Atomic, Inc., 33 University Road, Cambridge, Mass. 02138.

reading the clocks properly. For every revolution of the large hand (40 clock units), the small hand moves 45 degrees (5 clock units). One must learn to add mentally 40 clock units to the large-hand reading for every 45-degree travel of the small hand. The subsequent clock readings are then referred to a calibration curve or chart relating percent concentration to the clock reading. Misreading a clock by one complete revolution (40 clock units) is not unusual.

Many users go a step further by translating their working curve to a specially printed clock face bearing a spiral which is substituted for the original clock face. The reading of the percent concentration is then taken directly from the spiral where the large hand stops. Since the available spirals cover more than one revolution, it is still necessary to continue to note carefully the position of the small hand to eliminate misreading the correct percent.

Without using background correction, not much is gained by extending curves for more than two revolutions. (This is four RC time constants for the usual 5-sec RC discharge in the Baird Atomic System.) However, after installation of an automatic dynamic background system on our equipment,[2] it became possible to extend curves to three revolutions (or 6 RC's). This produced a near chaotic situation for the instrument operators and we found an increase in the number of misreadings due to misjudgment of the angle of the small hand.

We believe our solution to this problem is unique in its simplicity and application, although we have learned of other unpublished solutions (Veeder counters or colored lights to indicate full revolutions of the large hand).

We reasoned that by replacing the large hand with a clear plastic disk (0.060 in. thick), we could cause the spiral (containing the percent calibration curve) to

2. W. R. Kennedy, Appl. Spectry. 19, 74 (1965).

rotate. In addition, by replacing the small hand with another spiral hand whose outward progression is eight times that of the calibration spiral, the two spirals would always cross on a vertical line.

In essence, the small hand spiral follows the continuously rotating calibration spiral for the whole length of the curve with no ambiguity; nor is there any need to know how many revolutions the disk has made. The difficulty which might arise from using this system to revert to clock readings, can be overcome by using different colored segments for the disk spiral. Each color would indicate a different revolution and the small spiral hand position, relative to the color, would indicate which multiple of 40 clock units to add to the clock reading of the disk.

The plastic disks are made by photographing the available paper faces containing the spiral. The negative is reversed in enlarging, and printed on commercially available emulsion-coated plastic.[3] After processing, the disks are sawed out and calibrations scratched onto the spiral portion. The scratches are filled in with china-marking pencils (two calibrations can be accomodated) and sprayed with clear plastic. A small hole is drilled in the center of the disk for mounting onto the clock shaft. The small spiral hand was reproduced graphically and sawed from another piece of plastic and painted black.

To assemble, the clock hands and face are removed. The clock face is replaced with a white piece of paper using the usual method of attachment (glueing to the reversed metal clock face and fastening with the three screws provided). A vertical black line is drawn on the white paper. Next, the small hand spiral is placed on the outer (larger) clock shaft and secured with the usual nut provided. Finally, the plastic disk is mounted on the clock shaft making sure its "zero" is aligned

3. "Photoplast Plates," Eastman Kodak Company, Rochester, New York.

with the vertical line drawn on the white backing paper. This is best done by taking a spare clock shaft with square end, heating it and pressing into the back side of the plastic over the center hole so the disk will line up when mounted. There is no need to replace the glass face of the clock. However, if this is desired, first make sure the plastic disk is aligned so that it does not wobble.

Once assembled, the clock will operate in the normal manner, but with some additional side advantages. First, the percent readings are always upright and always taken on the vertical line for there is no need to search the clock face. Moreover, if calibrations need to be changed, it is only necessary to remove the disk and replace it with another. Previously the whole clock face had to be disassembled. We believe judicious use of a keyed magnetic attachment for the plastic disk could provide an almost instantaneous change over from one matrix to another. However, these are side advantages and were not envisioned in the original concept.

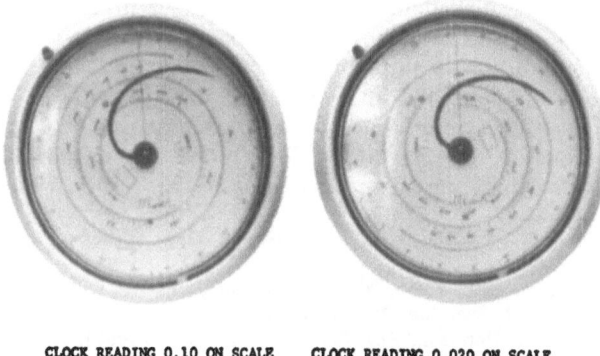

CLOCK READING 0.10 ON SCALE
(CLOCK UNITS = 45.1)

CLOCK READING 0.020 ON SCALE
(CLOCK UNITS = 67.5)

FIG. 1. Two clock readings using new plastic disk and spiral hand.

The disadvantage is that the plastic disk blanks, as well as the small spiral hands, must be hand made. We found the latter the more difficult because of the odd-shaped center hole.

Photographs of an assembled clock with two different readings are shown in Fig. 1.

1.22 An Improvement to the Weekley–Norris Electronic Computer

W. R. Kennedy

American Cast Iron Pipe Company, Birmingham, Alabama 35202

Weekley and Norris[1] have described an electronic computer adapted to a Baird–Dow[2] condenser discharge readout system.

Their Fig. 7 describes a servo-driven dual-tandem potentiometer, one winding of which is logarithmic and the other linear. The error signal balances their computing amplifier output (their Fig. 6) with the voltage picked off the log-wound potentiometer. At this point, the linear potentiometer indicates parts per million in millivolts direct.

Solid-state art has advanced significantly in the last few years, and it has been shown that the transfer function of a silicon planar transistor is a logarithmic impedance function.[3,4] The application of

1. R. E. Weekley and J. A. Norris, Appl. Spectry. **18**, 21 (1964).

2. J. L. Saunderson, V. J. Caldecourt, and E. W. Peterson, J. Opt. Soc. Am. **35**, 681 (1945).

3. W. L. Paterson, Rev. Sci. Instr. **34**, 1311 (1963).

4. J. F. Gibbons and H. S. Horn, IEEE Trans-Circuit Theory **11**, 49 (1964) (obtainable from Philbrick Researches, Inc., Dedham, Mass.).

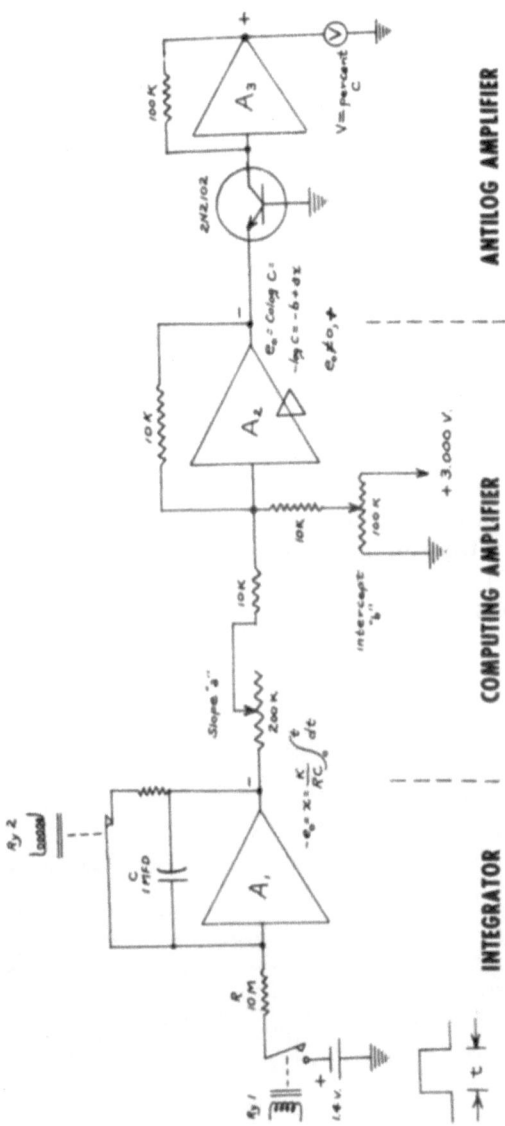

FIG. 1. Schematic diagram for a single-channel direct concentration computer.

another operational amplifier and logging transistor
to the output of the Weekley–Norris computing amplifier (their Fig. 6) will allow a voltmeter to read
directly in percent, provided the intercept voltage
''b'' is selected so that the computing amplifier output is the colog rather than the log, as in their application.

The finalized computer incorporating the integrator
(Weekley–Norris Fig. 5), the computing amplifier
modified to give colog output, and the logging transistor replacement circuit for their servo (their Fig.
7), is shown here in Fig. 1.

Slopes and intercepts can still be selected with
potentiometer settings as with their application. The
transistor shown (2N2102) can log seven decades.
Component constants given here are commensurate
with (1) a 30-sec instrument readout time (which is
also the integrator pulse-duration time controlled by
R_yl), (2) transistor input logging range, and (3)
operational-amplifier restrictions. The actual slope
R_f/R_i worked out to be $\frac{1}{12}$ to match the 5-sec RC

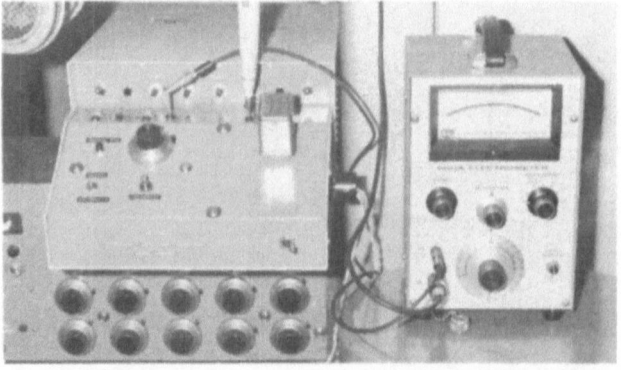

FIG. 2. Prototype model of the single-channel direct concentration computer.

time constant of the Baird installation at Acipco. This was sufficient for over four decades.

A prototype model was built (Fig. 2) to test on one channel of the Baird instrument and performed very satisfactorily on a background-corrected chromium curve from 4.5% (45 V) to 0.03% (0.3 V). In actual application for production purposes, the transistor would need to be placed in a constant-temperature oven or be temperature compensated.

Obviously, this improvement removes the electromechanical servo and its inherent problems, and also eliminates the specially wound potentiometer it demanded.

Semiconductor Current Regulator for the 1.23
Spectrographic Arc*

H. H. Conover, J. T. Peters, and M. Lalevic

Drexel Institute of Technology, Philadelphia, Pa. 19104

An inexpensive semiconductor-controlled regulator for the dc spectrographic arc is described. It is used for maintaining steady arc currents from 1 to 15 A, ±50 mA with a 120-V dc source. It is unique in that it employs solid-state circuitry. As compared to the conventional multipurpose power source with current regulators, the present model offers advantages of compactness, easier installation, and economy.

This current regulating device was designed for use in qualitative spectrographic analysis, as well as

---

*This was a research project assigned to H. H. Conover, physics major, B.S. June 1964. It grew out of In-House Project No. 7, now No. 21.

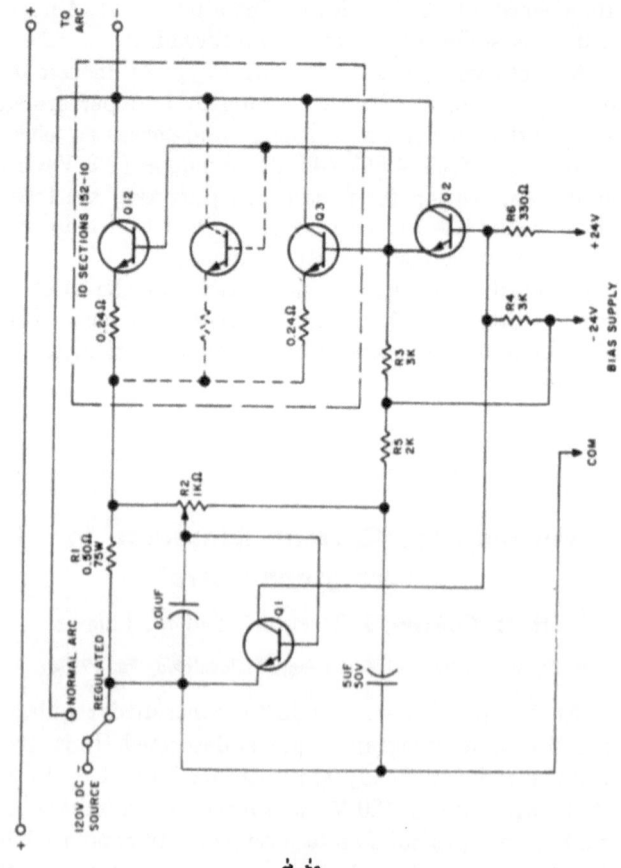

FIG. 1. Circuit dia-
gram for current reg-
ulator.

quantitative, employing a 120-V dc-arc source. It differs from the usual current regulators in that it makes use of high-wattage transistors. The instrument is capable of handling currents from 1 to 15 A with 50 mA stability.

It could be greatly improved through use of a zener reference diode, but this was not felt to be warranted in this particular application. Power or energy dissipation was achieved by water cooling.

Westinghouse type 152–10 silicon power transistors, N–P–N, were used throughout, having been found to offer typically high h_{FE} of 50 with conservatively rated power capability and collector–emitter voltage. The current diagram is given in Fig. 1.

Current sampling is provided through [R–1] (the numbers in brackets correspond to those in Fig. 1). Operating current is established through a balance of voltage appearing on the tap of [R–2] and V_{EB} of [Q–1]. [Q–2] serves as a current amplifier and drives the bases of the regulator bank comprised of [Q–3] through [Q–12].

Biasing is such that, prior to striking the arc, [Q–2] is conditioned for 25-A initial current through the regulator bank. Bias voltages are provided by internal, zener-stabilized, plus and minus 24 V supplies.

All transistors, except [Q–1], were mounted directly in contact with a water–cooled copper heat sink. Water cooling was supplied by a single loop of $\frac{1}{4}$-in.-i.d. copper tubing affixed by soft solder to the heat sink. Water connections to the heat sink were made by plastic tubing. Electrical conduction through the water was negligible at all voltages used.

An ammeter was used to check on regulator activity. It recorded the base current of the regulator
bank. Readings of arc current and voltage were obtained from external meters. Relay switching is provided in the instrument so that it may be used with
or without the current regulator. (See Fig. 2).

The current regulator is connected in series with
the conventional-arc resistance bank. For operation
without regulation, a relay contact is closed across
the regulator bank and sensing resistor. But, for
currently-regulated operation, the resistance bank is
set to zero or minimum resistance; the electrodes are
brought into contact, and the desired current obtained by adjusting the regulator. The regulator then
retains this value for current when the electrodes are
separated to initiate the arc. Regulation of the current at this value continues until the voltage drop

FIG. 2. Photograph of current regulator, showing transistors
and heat sink.

across the arc gap is sufficient to cause saturation of the regulator bank.

The instrument has been successfully used in Drexel's spectroscopy laboratory. Operational currents used were 10 to 15 A, although satisfactory operation may be had through the 1 to 15 A range of the instrument. Very little current variation was observed, on a conventional 0 to 15 A meter, from shorted electrodes to arc lengths of several centimeters. Rapid fluctuations of current that are so often encountered when arcing certain types of samples are eliminated by this regulator.

Acknowledgment. The authors wish to express their appreciation for the cooperation of Drs. F. K. Davis, G. S. Sasin, and E. J. Rosenbaum.

Punched Cards as Position Indicators for Spectrographic Plates 1.24

J. M. McCrea

Applied Research Laboratory, United States Steel Corporation, Monroeville, Pennsylvania 15146

Rapid location of the significant regions on spectrographic plates by use of templates or position indicators can speed up plate measurements considerably. Punch cards for computer use provide a basic grid of 960 positions over a 7.0-in. ×2.9-in. area, and in several designs the 12 rows and 80 columns are legibly numbered. Holes in a punch card serve not only to indicate position, but can also be used as windows for photometric measurements if the card is used as a mask directly on the uncoated side of the plate.

The author has used evenly spaced columns of holes in a punch card in studies of processing fog distribution over very sensitive emulsions such as the Ilford Q2 and Kodak types I and SWR. Transmittance data were taken across seven rows as a function of column displacement on 2-in. wide test plates. For clarity of illustration, Fig. 1 shows an unusual case in which the processing fog level varied essentially monotonically with position on the plate and the curves for the alternate rows shown lie clearly separated on a plot. A contour diagram for the plate showing processing fog distribution could be prepared from data interpolated on Fig. 1.

This spectroscopic trick with punched cards has proved useful for obtaining data on processing fog level. In cases involving specified lines rather than arbitrary background areas, it would often be neces-

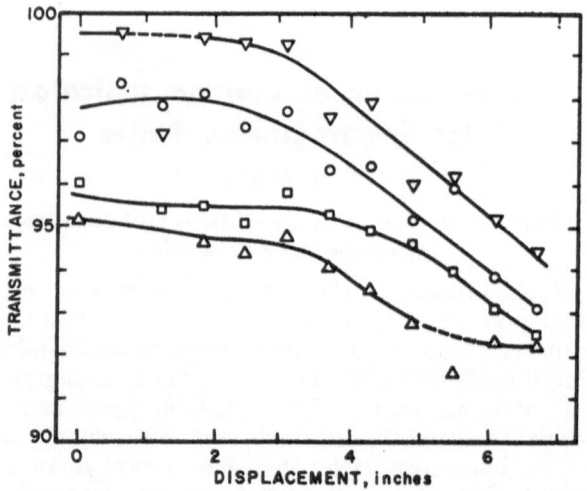

Fig. 1. Variation of transmittance with displacement along rows of a punch-hole mask for a sensitive plate.

sary to remove the region between two adjacent key punch holes to prepare a suitable position indicator from a punch card.

Derivative Scanning for Wavelength Identification 1.25

J. P. Walters* and H. V. Malmstadt

*Department of Chemistry and Chemical Engineering,
The University of Illinois, Urbana, Illinois*

With spectrographs of low dispersion, wavelength identification can become difficult even under high magnification. With prism instruments, such as the Bausch and Lomb Medium Quartz spectrograph used here, the problem is further compounded by nonlinear display on the photograph. However, if a small region of the spectrum is recorded, it is possible to calculate a dispersion that will be approximately linear. The problem is then reduced to measuring the distance between lines of known wavelengths on the strip chart recording. Such distances can be measured with good accuracy if a first derivative tracing is made from the densitometer output, rather than the conventional transmission recording.

The densitometer used here is a National Spectrographic Laboratories, model TM-102, ''Spec-Reader'' unit. It is equipped with an output jack for recording, plus digital readout of transmission values. The technique for wavelength identification is as follows. For purposes of orientation and preliminary identifications, an iron-arc reference spectrum is placed imme-

*Present address: University of Wisconsin, Madison, Wisc.

diately below the test spectrum, in this case a spark spectrum of aluminum, with the aid of the Hartmann diaphragm on the Medium Quartz instrument. The plate is placed in the densitometer such that part of the densitometer slit covers the iron reference spec-

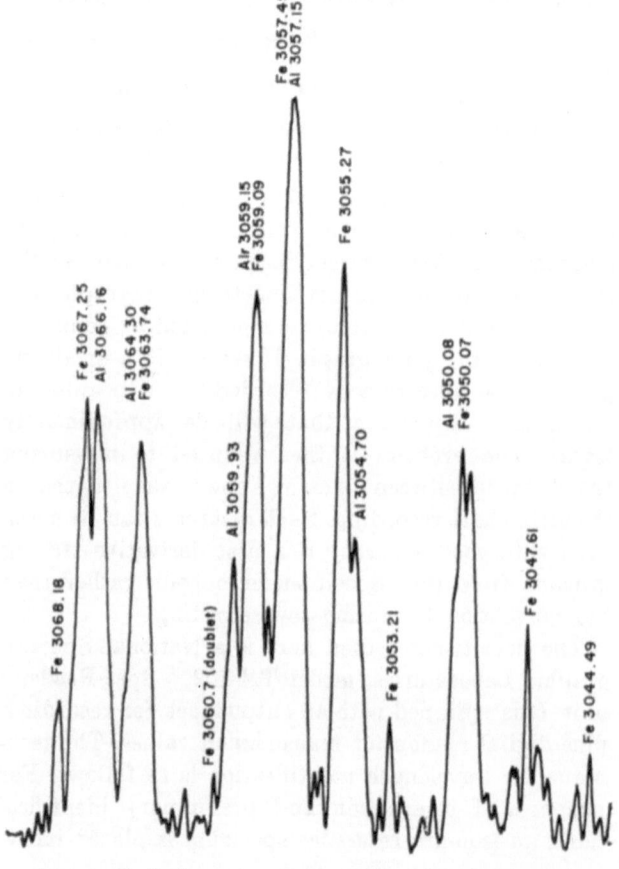

Fig. 1. Preliminary composite tracing for orientation.

trum, while the remainder covers the spark spectrum. The composite tracing shown in Fig. 1 is then made. By adjusting the position of the plate, the relative height of the iron lines compared to the spark lines can be varied. Known wavelengths are marked on the tracing.

A first derivative scan is then made over the spark spectrum for the same wavelength region as in the composite tracing. Here, a Heathkit, model EUW-20A, servo recorder is used for both scans. The derivative tracing is obtained by taking the output of the densitometer amplifier through an RC differentiator whose time constant has been optimized for a particular scan speed. The output of the differentiator is taken directly to the Heathkit recorder.

Figure 2 shows normal and derivative scans obtained by the above method. The first derivative scan of a line closely resembles the second derivative of a sigmoidal curve. The maximum height of a line (minimum transmission) corresponds to the zero level on the derivative tracing. Thus, location of the peaks or even badly skewed spark lines is readily accomplished.

The zero points on the derivative scan are marked; known wavelengths are obtained from a composite scan such as shown in Fig. 1. Guide lines are drawn through the zero points, perpendicular to the scan axis. The distances between guide lines are measured with drafting dividers and a high-quality machinist's scale ruled in millimeters. If there are at least three known lines in the vicinity of the unknown line, covering not more than 20 Å total scan distance, it is possible to assume linear wavelength display. From the measured dispersion, interpolation to the unknown line serves to pinpoint its wavelength with sufficient

accuracy to allow positive identification from wavelength tables.

The accuracy of the above method depends on the
linearity with which a scan can be recorded. Such
factors as nonuniform recorder chart speed, twisting
or stretching of the recorder chart paper, nonuniform

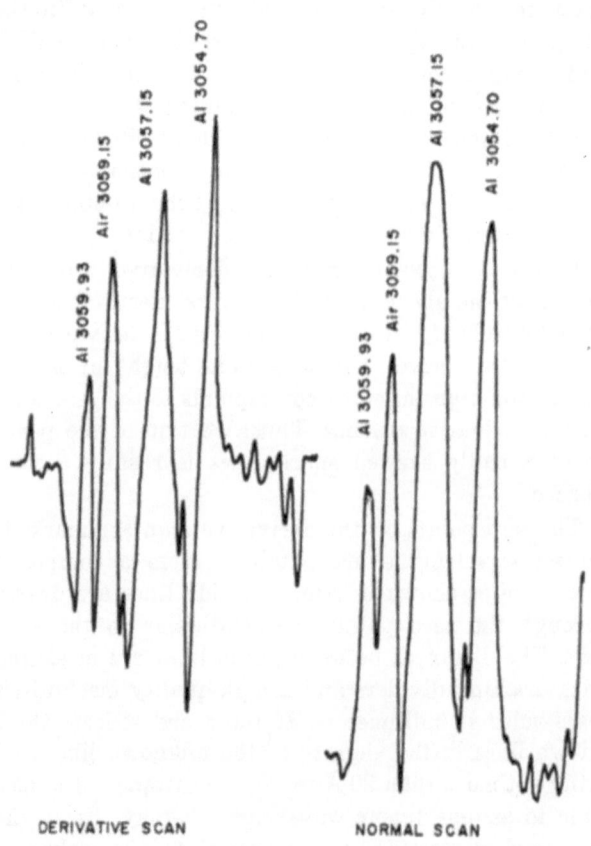

FIG. 2. Normal and first derivative tracings.

Table I. Determination of displayed dispersion.

Ref. line in Å	Test line in Å	Wave-length difference in Å	Distance in mm	Displayed dispersion in Å/mm
3068.18	3067.25	0.93	5.7	0.169
3068.18	3067.25	0.93	5.5	0.163
3067.25	3066.49	0.76	5.7	0.133
3067.25	3066.49	0.76	5.8	0.131
3066.49	3063.94	2.35	16.2	0.145
3066.49	3063.94	2.35	16.2	0.145
3063.94	3060.99	2.95	22.3	0.133
3063.94	3060.99	2.95	22.6	0.131
3060.99	3059.09	1.90	13.3	0.143
3060.99	3059.09	1.90	13.0	0.146
3059.09	3057.45	1.64	11.7	0.140
3059.09	3057.45	1.64	11.7	0.140
3057.45	3055.27	2.18	15.3	0.142
3057.45	3055.27	2.18	15.5	0.141
3055.27	3055.21	2.06	14.4	0.143
3055.27	3055.21	2.06	14.6	0.141
3053.21	3048.47	4.74	34.3	0.138
3053.21	3048.47	4.74	34.5	0.137
3068.18	3048.47	19.71	138.5	0.142
3068.18	3048.47	19.71	138.4	0.142
Average				0.141
Average deviation				0.006

densitometer scan speed, and emulsion stretching must be taken into account. As a test, two derivative scans were made over the wavelength region shown in Fig. 1 for two iron-arc spectra. The displayed dispersion was then determined between pairs of iron lines over the entire range, plus between the two extreme iron lines. The results are shown in Table I. It is clear that essentially identical values of the displayed or recorded dispersion are obtained for

both spectra. From data such as this, the wavelength
of an unknown line can be determined to within
0.04 Å, an uncertainty smaller than the half-intensity
breadth of most lines observed.

1.26   Device for Measuring Spectral Linewidths*

Paul R. Barnett

U. S. Geological Survey, Denver, Colorado 80225

Conventional methods of microphotometry measure
the portion of an incident beam of light that is transmitted through a given spectral line on a photographic film or plate. This percentage transmittance
is, in effect, a measure of the density of the line; for
density is equal to the logarithm of the reciprocal
of the transmittance. These methods are well founded
and are not likely to be replaced for those densities
on or near the linear portion of the H and D curve.
However, as the density approaches the photographic
saturation level, the methods become less reliable and
eventually fail. The chief causes for this failure are
(1) the diminishing efficiency of the photographic
emulsion when most of the silver halide has been exposed, and (2) the lack of reliability of the microphotometer when the transmittance is below about
3%.

Many methods have been employed to reduce the
density to usable values. They include diluting the
sample, using step filters, reducing the exposure of
the whole spectrogram, or placing a filter at the plate
in order to reduce the exposure of a limited wavelength region—all of which require expenditures of
additional time and/or materials.

*Publication authorized by Director, U. S. Geological Survey.

Another photographic phenomenon that bears the same relationship to incident light as does density is broadening of the image. This broadening is due to scattering of the light by the silver halide grains into the shaded area adjacent to the image.[1] The more intense the incident light, the greater is the scattering and, hence, the broader the photographic image. The width of the spectral line in spectrography is a reliable measure of the amount of radiant energy producing the line. The width is, in fact, directly proportional to the logarithm of the integrated incident light.

In an excellent paper on the use of line width as an alternative for density measurement in spectrochemical analysis, Junkes and Salpeter[2] point out that astronomers had used the principle of image broadening for estimating stellar brightness for half a century before it began to be used in spectrography. Eisenlohr and Alexy[3] in 1937 first demonstrated its usefulness for concentration measurements in spectrochemistry. Although this innovation was a significant milestone in spectrographic analysis, the linewidth method has been slow in coming into general use during the ensuing years.

The chief reason for this slow development has been the lack of a technique or an instrument for rapidly and accurately determining the width of the line. Early workers[4,5] measured the width of the recorder

1. C. E. K. Mees, *The Theory of the Photographic Process* (The Macmillan Company, New York, 1964), Chap. 25, p. 1001.
2. J. Junkes and E. W. Salpeter, Ric. Spettroscopiche, Lab. Astrofis. Specola Vaticana **2**, (5), 205 (1958).
3. F. Eisenlohr and K. Alexy, Z. Physik. Chem. (Leipzig) A **179**, 241 (1937).
4. P. Coheur, J. Opt. Soc. Am. **36**, 498 (1946).
5. D. J. Hunt and D. L. Timma, "Studies of the Line Width Method of Spectrochemical Analysis," 4th Annual Pittsburgh Conference on Analytical Chemistry and Applied Spectroscopy (1953).

tracing of a line at some more or less arbitrary base (percent transmission). This technique is good in theory but is inaccurate in practice because of the inconstant rate of movement of the plate carriage on most recording microphotometers and, hence, poor precision in the width of the tracing.

Junkes and Salpeter[2] designed a variable slit of large aperture for a microphotometer and measured "effective linewidths" in the following way. With the slit set at narrow opening, the transmittance of a clear portion of the plate is observed. With the line placed in front of the slit, the slit is opened until the transmittance is the same as for the narrow slit on the clear plate. The new opening is called the effective width of the line. This method has been shown to work but requires the construction of a somewhat complicated variable slit and the modification of the microphotometer to which it is affixed.

Bastron, Barnett, and Murata[6] clocked with a stopwatch the time required for the microphotometer to scan the line between 5% transmission values on opposite sides of the line. This technique works well for microphotometers with a light-weight, moving slit pulled across the line by a constant-speed motor. The method will not work on most microphotometers where the slit remains stationary and the carriage must move to scan the line. This is due to the inconstant rate at which a heavy carriage is moved by an underpowered motor and a long mechanical linkage.

Figure 1 is a top view of a device that may be attached to a microphotometer of the moving-carriage type to convert it into an instrument for the accurate and relatively rapid measurement of linewidths. The device is essentially a micrometer depth gauge (a) with a 4-in. base (b), a 6-in rod (c), and a run of 1 in. The gauge is secured to the end of the plate car-

6. H. Bastron, P. R. Barnett, and K. J. Murata, U. S. Geol. Surv. Bull. **1084-G**, 165 (1960).

FIG. 1. Top view of the device. (a) Micrometer depth gauge, (b) 4-in. base, (c) 6-in. rod, (d,d') stainless-steel blocks, (e) aluminum wheel, (f) millimeter scale.

riage as follows: two $\frac{3}{4}$ in. $\times \frac{3}{4}$ in. $\times \frac{5}{16}$ in. stainless-steel blocks (d,d') are fastened to the bed of the plate carriage with two Allen screws countersunk in each block. Holes are drilled through the narrow portion of the ends of the base of the gauge and the gauge screwed to the steel blocks through these holes. The geometry is such that the projection of the plane of the lower surface of the spectrographic plate is tangent to the gauge rod at the lower end of its vertical diameter. This prevents the rod from hitting the rhomb (or slit) of the microphotometer and possibly damaging it.

In order that small movements of the depth gauge may be read more accurately, an aluminum wheel (e) approximately 4 in. in diam is friction fitted around the knurled thimble of the depth gauge. The wheel has a $\frac{1}{2}$-in. flange and a hub to accommodate a set screw. A millimeter scale (f) is fastened to the circumfer-

ence of the wheel. An index made of brass extends
out from the near end of the base of the depth gauge.

As the wheel is turned to the right, the rod of the
gauge pushes the spectrographic plate to the left with
respect to the plate carriage. When the wheel is rotated to the left, a leaf spring (Fig. 2) attached to
the ways of the plate carriage with a thumbscrew
pushes the plate back to the right. By reading the
wheel scale to the nearest $\frac{1}{2}$ mm, movements of the
plate can be read to 0.001 mm, or 1 μ, for the device
with a micrometer screw of 1/40-in. pitch and a
wheel with a circumference of 317.5 mm.

To use the device in connection with the microphotometer in order to measure the width of a line,
the plate carriage is moved until the line is just to
the right of the light beam entering the pickup slit.
The plate carriage is then secured so that it will not
move. This can be done by putting small but powerful
magnets on the steel top of the microphotometer and
against the ends of the plate carriage. To always
have the plate carriage so positioned that neither

FIG. 2. Leaf spring for returning plate.

end extends beyond the top of the microphotometer, it is sometimes necessary to insert a rectangle of aluminum with dimensions approximately the same as a half-plate between the rod of the depth gauge and the plate being read.

With the carriage secured, the depth gauge is rotated to the right until the microphotometer reads 5%, or some other preselected value. At this position the scale on the wheel is read. The gauge is then advanced, moving the line across the slit of the microphotometer, with the percent transmission during the operation being less than the preselected value. The gauge movement is stopped when the percent transmission again reaches the preselected value, and the wheel scale is again read. The difference between the two scale readings is proportional to the width of the line.

This device is being used in our laboratory for routine analysis of elements whose best lines are too dark for density measurements. Although not as rapid a procedure as densitometry, it is less time consuming than redoing the sample at a lesser exposure or greater dilution.

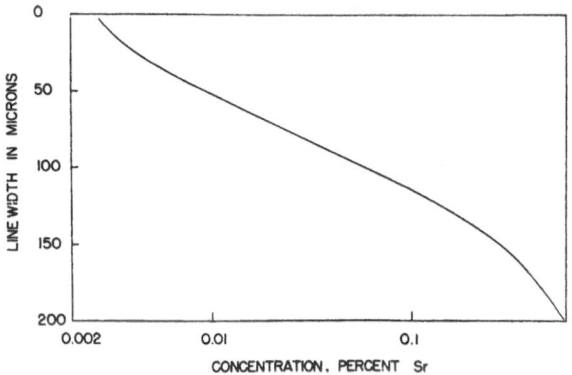

Fig. 3. Working curve for strontium (4607-Å line).

Standards are always exposed on the same plate with the unknowns. Working curves are drawn by plotting the width of the line against the log concentration. The curves are typically S-shaped with a nearly linear central portion (Fig. 3). The standard deviation for repeat line-width measurements has been determined to be about 0.8 μ. This represents a coefficient of variation of 2% for a line 40 μ wide and $\frac{1}{2}$% for a line 160 μ wide.

The device is inexpensive, is easy to fabricate and attach to the microphotometer, does not impair the normal functions of the microphotometer, and can be quickly detached. No great skill is required for its operation.

SECTION 2
INFRARED SPECTROSCOPY

Protective Holder for KBr Pressed Disks 2.1
Containing Reactive Samples*

P. A. Staats and H. W. Morgan

Oak Ridge National Laboratory, Oak Ridge, Tennessee

Two basic approaches have been recommended for
the preparation and handling of KBr disks containing chemically reactive samples. Several workers have
suggested a "sandwich" technique,[1,2] in which the
sample is contained in the center of the disk, protected by outer layers of KBr. Since water does diffuse into the pressed disk,[3] this delays but does not
prevent reaction. Price and Maurer[4] have described a
method in which the sample is uniformly distributed
through the KBr, and the surfaces protected by
mounting the disk between alkali halide windows in
a conventional demountable cell. While this method

*Research sponsored by the U. S. Atomic Energy Commission
under contract with Union Carbide Corporation.

1. E. G. Brame, Jr., S. Cohen, J. L. Margrave, and V. W.
   Meloche, J. Inorg. Nucl. Chem. **4**, 90 (1957).
2. E. W. Lawless, Anal. Letters **1**, 153 (1967).
3. R. A. Buchanan and W. A. Bowen, Jr., J. Chem. Phys.
   34, 348 (1961).
4. W. H. Price and R. H. Maurer, Appl. Spectry. **17**, 106
   (1963).

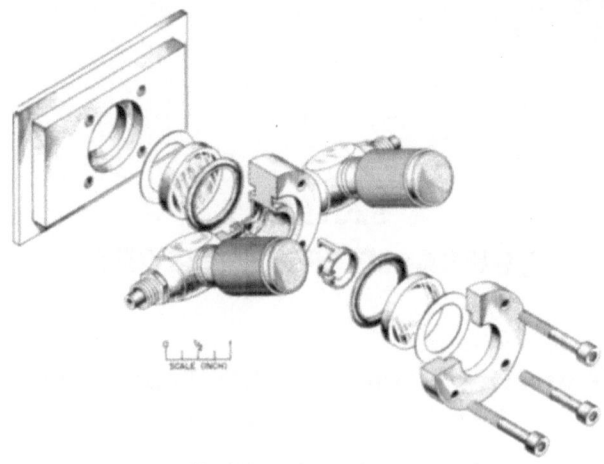

Fig. 1. Exploded view of pellet holder.

provides some protection against atmospheric water
and oxygen, it is frequently necessary to have more
rigorous control of the sample environment. The sample holder described here allows the KBr pellet to
be held in vacuum, under an inert atmosphere, or
in contact with reactive gases in a static or flow system. The reactivity of the sample and any changes
in spectra may be easily determined.

Construction details are shown in the exploded
drawing of Fig. 1. The holder was designed for a
13-mm diam disk, but the basic design can be used
for disks or samples of any appropriate size or shape.
The windows are 1 in. in diam, sealed with O-rings.
The thickness of the cell is determined primarily by
the valves used. The Hoke type A-431 valves used
here and attached as shown allowed a thickness of
approximately $\frac{1}{2}$ in. The male end connections on the
valves were altered by removing the threads from
one end for soldering to the center section. The other
end was completely removed and replaced with a

connector for polyethylene tubing. These alterations are optional. The holes connecting the disk cavity to the valves are drilled so that both sides of the disk are exposed to gas flowing through the holder. The disk is held in place by a small spring clip.

The KBr disks are transferred as rapidly as possible from the die into the holder (in a drybox, if one is available), and the cell assembled. The holder has been used in this laboratory for several years for reactive samples. A convenient source of an inert atmosphere has been the off-gas flow from a Dewar filled with liquid nitrogen (and connected to the inlet valve by a length of tubing).

New Technique for the Preparation of 2.2
KBr Pellets from Microsamples

Harry R. Garner and Herbert Packer

*The Harshaw Chemical Company, Division of Kewanee Oil Co.,
1945 East 97th Street, Cleveland, Ohio 44106*

The most popular techniques for collection of microsamples on KBr powder prior to infrared spectroscopic examination are evaporation from a solution of the sample in the presence of KBr powder and lyophilization.[1]

Many microsamples originate from thin-layer chro-
the area of the chromatogram containing the sepa-
matographic (TLC) separations. The sample is
separated by TLC, the adsorbent removed from
rated material, the sample eluted with a suitable solvent, filtered to remove the suspended adsorbent, and

1. W. B. Mason, Pitts. Conf. Anal. Chem. and Appl. Spectry., March 1958.

finally mixed with KBr by one of the above techniques.

By using a porous triangle of pressed KBr, a Wick-Stick™, in a small glass vial capped so that evaporation is restricted to the center of the vial, the filtration of adsorbent and deposition of the sample on KBr can be accomplished in a single step. The adsorbent containing the sample is scraped from the thin-layer chromatographic plate and transferred to a glass vial containing a Wick-Stick using a thin-stemmed funnel to prevent the adsorbent from dusting on the top half of the Wick-Stick. A suitable eluting solvent is added and a vented cap is placed on the vial.

Figure 1 shows the Wick-Stick apparatus with a sample of dye partially collected from silica gel G using ethyl ether as the eluting solvent. The pressed KBr triangle is 2.5 cm high, 0.8 cm wide at the base, and 0.2 cm thick. The glass vial is 3.5 cm high with a 1.0-cm inside diameter. The vent hole in the stainless steel cap is 0.3 cm in diameter. A stainless steel spring

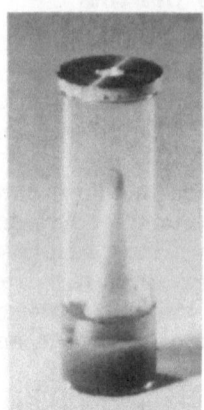

Fig. 1. Wick-Stick apparatus. A sample of dye is shown partially collected from silica gel G.

clip holds the Wick-Stick upright and centered in the vial.

The solvent climbs the Wick-Stick by capillary action and evaporation takes place preferentially at the apex of the KBr triangle depositing the sample. The adsorbent is effectively filtered by the porous KBr. Emission spectrographic analysis has shown that less than 10-ppm adsorbent is detected in the tip of the Wick-Stick after elution from either alumina gel G or silica gel G.

One to two mm of the tip is cut off with a sharp scalpel, mashed on a clean metal surface, and pressed into a transparent disk using a microdie. Most of the sample is concentrated on a quantity of KBr which is slightly in excess of the amount required to press 1.5-mm microdisks. For ultimate sensitivity, the KBr can be carefully scraped from the edges and surface of the Wick-Stick tip and incorporated into a 0.5-mm microdisk.

Evaporation of the solvent can be accelerated by maintaining the temperature $10°–20°C$ below the boiling point of the solvent and directing an air jet across the cap of the vial. The vial holds 1–1.5 ml of solvent which will evaporate under these conditions in less than 1 h. Two or three solvent passes effectively concentrate the sample at the tip. While concentration may require a half-day, the operator does not have to attend the samples during this time and is free to perform other duties.

Satisfactory spectra for qualitative analysis are obtained from 10–50 μg of sample when a 1.5-mm-diam micro-KBr-die and a beam-condensing unit are used.

Figure 2 shows the spectrum obtained from 20 μg of pyrogallol collected from an ethyl ether solution and Fig. 3 presents the spectrum of 50 μg of cholesterol which was separated from a steroid mixture by TLC and collected by the Wick-Stick method.

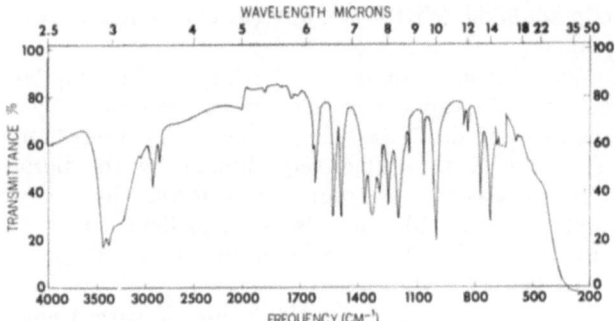

FIG. 2. 20 μg of pyrogallol collected on a Wick-Stick from ether solution.

The scan speed of the infrared spectrophotometer was adjusted to provide the entire spectrum in 15 min.

Wick-Sticks are applicable to the collection of samples for macropellets, however, the amount of sample required to provide adequate sensitivity increases in proportion to the amount of KBr in the pellet. An instrument which incorporates provision for expanded scale operation will minimize the sample-size requirements for both micro- and macropellets.

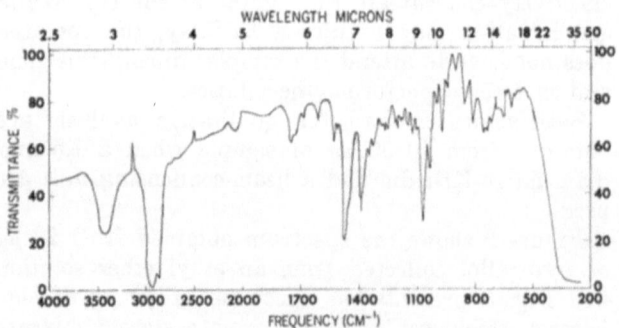

FIG. 3. 50 μg of cholesterol separated from steroid mixture by TLC and collected on a Wick-Stick.

A Technique for the Use of the "Mini-Press" in Micro Infrared Spectroscopy   2.3

Leo A. Bauman, Jr.

Food and Drug Administration, Buffalo, New York 14202

The Wilks Mini-Press has been used extensively in our laboratory for detecting microgram quantities of various pharmaceutical drug products. Their instruction sheet[1] states: "grind approximately 0.5 to 1.0 mg solid sample." However, by using the method described below we have been able to obtain spectra showing strong absorption bands of various pharmaceutical drugs with as little as 20 μg of material (see Fig. 1).

Procedure: Weigh 65–70 mg of KBr, place in Mini-Press, and apply sufficient pressure to produce a transparent KBr disk. After removing the bolts, punch out *only* the window of the disc into a suitable mortar, leaving intact the KBr in the threads. Add about 50 to 100 μg of sample to the mortar and intimately grind the KBr disk with the sample. Replace one bolt snug to the KBr still present in the threads. Transfer the sample matrix to the press and form a clear pellet in the usual manner. After removing the bolts, place the barrel in the sample holder.

Rotate the barrel containing the disk in the sample beam of the instrument in a transparent region of the sample (3000–2500 cm^{-1}) in order to obtain maximum transmittance. Attenuate the reference beam to position the pen to about 90% transmittance.

Adjust the gain and balance with the sample in position at the same 3000–2500 cm^{-1} transparent region. Scan spectrum.

---

1. Wilks Scientific Corp., "Instruction for Operation and Maintenance, Mini-Press®" (10 31 66).

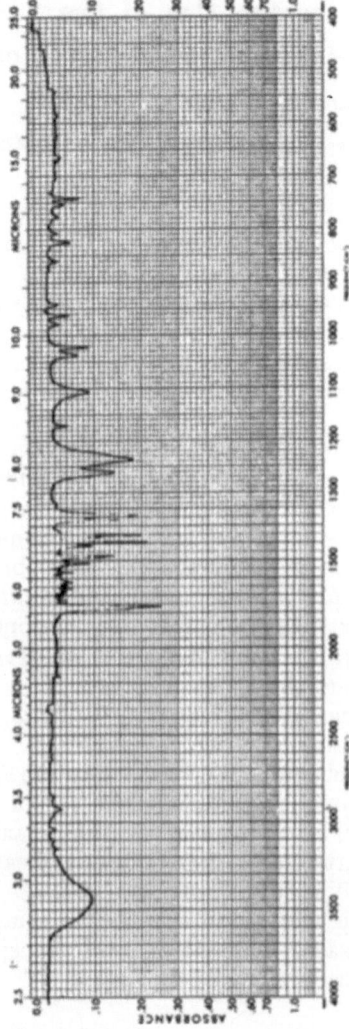

Fig. 1. Infrared spectrum of 20 μg of furazolidone, prepared by using the Mini-Press technique.

Discussion: This procedure places a maximum amount of compound in the energy beam. The KBr matrix in the threads, which is not in the path of the beam, does not contain any of the compound. Because all of the compound is placed in the path of the beam this technique yields a spectrum of maximum intensity per given weight of compound.

Use of less than 60 mg of KBr results in a disk which does not remain in the barrel. For the same reason a quantitative transfer of the sample matrix to the barrel is necessary. After use, all of the KBr should be washed from the barrel in order to prevent corrosion.

Laboratory Preparation of Polyethylene Pellets   2.4

W. B. Barish, G. T. Behnke, and K. Nakamoto

Department of Chemistry, Illinois Institute of Technology, Chicago, Illinois 60616

The development and increasing application of far-infrared spectroscopy has led to a corresponding need of sample-preparation techniques suitable for the far-infrared frequency range. With respect to solid-state samples, methods used for normal-infrared range spectra have been applied with some success to the far infrared. Thus, samples may be mulled with the usual mineral oils since such oils have good transmission properties throughout the far-infrared range. Pellets made of cesium iodide transmit very well down to 175 cm^{-1}, and the method of preparation is essentially the same as for potassium bromide pellets. Interest in far-infrared spectra, however, often extends well below this frequency. A remarkably convenient matrix mate-

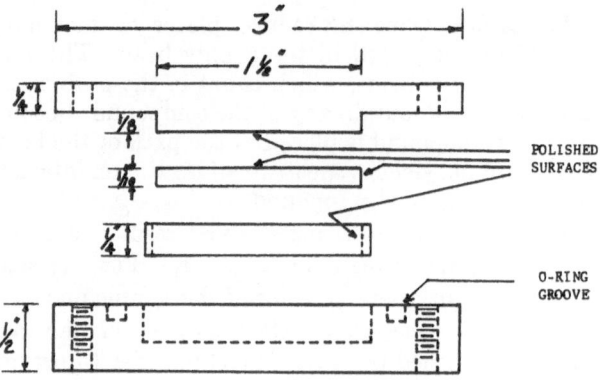

FIG 1. Polyethylene pellet die. Exploded, edge-wise projection view.

rial with excellent transmission from 650 cm^{-1} to at least 33 cm^{-1} (low-frequency limit on the Beckman IR-11) is polyethylene. The polymer in powder form, to which has been added a small quantity of sample, can be melted at relatively mild temperatures and then recooled to form a pellet semitransparent in appearance. These pellets are tough and durable, do not absorb atmospheric moisture, and do not alter perceptibly even after long periods of time.

The cell or die used to make the pellets may be as simple as two cover glasses[1] or as sophisticated as imagination can devise—or resources allow! The unit currently used in our laboratory consists of four pieces, one of which is a base plate of about 3-in. diam by ½ in. thick (see Fig. 1). Its center is machined to receive a ring disk assembly, the ring fitting rather loosely in the base. The disk, 1½-in. diam, is machined to fit accurately inside the ring and its upper surface is highly polished. An O-ring groove circles the disk and ring emplacement. The assembly is completed by a

1. L. May and K. J. Schwing, Appl. Spectry. 17, 166 (1963).

top plate $\frac{1}{4}$ in. thick having a raised center section of a diameter matching that of the removable disk. This surface also is well polished. After assembly the parts are clamped together by a set of screws, located outside the O-ring groove. When assembled for use, the space (for the pellet) between the removable disk and matching surface on the top plate is about 0.025 in. ±0.01 in. depending on the pressure applied to the gasket.

A photograph of the 4-piece die is shown in Fig. 2. The tube projecting from the top plate, not shown in Fig. 1, provides a small opening from the pellet area to atmosphere and serves as an air vent. Air trapped in the pellet space might otherwise expand into the molten polyethylene causing bubbles to appear in the pellet. It is probably unnecessary when surfaces in contact with the pellet are adequately polished. Our die, which cost about $30.00, is made entirely of stainless steel but for many samples this is hardly necessary. We decided on a pellet diameter of $1\frac{1}{2}$ in. because the slits on the IR-11 spectrometer are nearly this long. The fused polymer adheres with surprising strength to unfinished surfaces, so that polishing of the disk and top plate is strongly recommended.

FIG. 2. Photograph of die.

To make a pellet, the sample, finely ground, is intimately mixed (vibrator or agate mortar) with powdered polyethylene and the mix spread more or less uniformly on the disk in the die base. The top plate is positioned, rotated a bit on the sample to form a uniform layer, then clamped tight. The assembly can be heated (100°–120°C) on a hot plate or simply dropped in boiling water for 4–5 min. *Note that neither a high pressure hydraulic press, vacuum, nor a separate heater mounted on the press is required, as in previous descriptions of pressed polyethylene disks.*[2] After cooling and opening the die, the pellet can be peeled easily from the smooth metal surface. The thick pellet has the advantage of being structurally strong and the sample is protected from atmospheric oxidation and moisture. Moreover, such pellets may be conveniently stored for a long period in a reference file without change in the quality of spectra. Should it prove necessary to reclaim the sample, the pellet may be dissolved in warm xylene and the sample separated by appropriate techniques.

The amount of sample and polyethylene used will depend on the purpose of the work, intensity of bands to be observed, pellet diameter, etc. Normally we use 30–100 mg sample and 400–800 mg polyethylene. A suitable matrix polymer is "Microthene R 500," a microfine polyethylene powder received from the U. S. Industrial Chemical Co. of Tuscola, Illinois.

As illustration of pellet spectra, a far-infrared spectrum of *cis*-$[Pt(NH_3)_2Cl_2]$ imbedded in a pellet is shown in Fig. 3. The base line is a spectrum of a blank polyethylene pellet. It is seen that the polyethylene pellet is transparent below 650 cm^{-1}. The weak broad absorptions around 550–600 cm^{-1} are due to the polymer. If one is investigating for infrared spectra of

2. B. Smethurst and D. Steele, Spectrochim. Acta **20**, 242 (1964).

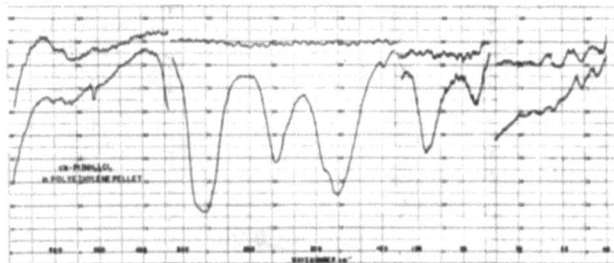

FIG. 3. Far-infrared spectrum of $Cis[Pt(NH_3)_2Cl_2]$ in polyethylene pellet. Lower line is the spectrum of a blank pellet.

samples sensitive to heat it is possible to form polyethylene pellets by high pressure instead of heat.[3,4] Highly crystalline, low molecular weight polymer is needed.

Acknowledgments—The authors wish to thank H. Hertrich for making the die. The Beckman IR-11 far-infrared spectrophotometer used in this investigation was purchased by the National Science Foundation Research Instrument Grant (GP-5417).

Preparation and Use of Powdered Silver Chloride as Infrared Matrix

2.5

T. M. Spittler* and Bruno Jaselskis

Department of Chemistry, Loyola University, Chicago, Illinois

Recently, Metzler[1] described a technique for preparation of inexpensive silver chloride disks for use in

3. C. Schiele and K. Halfor, Appl. Spectry. **19**, 163 (1965).
4. C. Schiele and F. J. Meyer, Spectrochim. Acta **22**, 1967 (1966).

*On leave of absence from Department of Chemistry, University of Detroit

1. R. K. Metzler, Anal. Chem. **36**, 2378 (1964).

running infrared spectra of aqueous samples. There are problems involved in preparing finely divided silver chloride, and little or no use has been made of this material as an infrared matrix. Because of the several disadvantages to the standard potassium bromide pelleting technique—hygroscopicity, fragility of pellets, possible reactivity of potassium bromide with the sample, and poor storage characteristics—we have found it useful to prepare pellets from powdered silver chloride. The presence of moisture and reactivity of the sample with matrix make potassium bromide undesirable, and an alternate matrix for pelleting samples would be highly advantageous.

Metzler's article suggests use of reagent grade silver chloride.[1] This material is contaminated and discolored. The contamination and discoloration can be avoided by using freshly prepared silver chloride. A fresh silver chloride pellet produced a flat trace in region 2–30 μ. This region is easily extended to 40 μ by using reference pellet of approximately the same weight (thickness).

Silver chloride is precipitated from a 2 M solution of reagent grade silver nitrate, with 2 M hydrochloric acid, working in a darkroom or dimly-lit room. After thorough washing of the precipitate with 2 M hydrochloric acid, it is air dried on the filter for about 15 min and then for several hours in Abderhalden pistol at 80°C. Then, a pellet is made from the dry coarse material to determine purity. The absence of the strong sharp nitrate bands at 13.4 μ and 12.5 μ indicates a high degree of spectral purity.

The coarse precipitate is placed in a thin-walled evaporating dish and covered with liquid nitrogen. When the sample and dish have reached equilibrium temperature, the silver chloride is ground to a fine powder by adding liquid nitrogen as needed to keep the sample covered. A porcelain or stone pestle must be used because a glass mortar and pestle cannot withstand the extreme thermal shock. This technique yields a finely divided powder.   (However, if the precipitate is entirely dry, the grinding procedure is considerably more difficult than otherwise.)

The powdered sample is returned to the drying pistol for another four to five hours to remove any water that may have condensed onto it at low temperature. The silver chloride is stored in containers that are light tight. Clear pellets of better than 80% transmittance have been prepared up to six weeks after final grinding.

The finely divided white powder seems to be non-hygroscopic and no precaution is taken to exclude moisture other than storage in closed vessels. The pellets are prepared and pressed using the conventional potassium bromide technique.

The above technique has many advantages over potassium bromide. There is essentially no moisture problem in achieving a uniform mixture of sample and matrix in the standard mixing capsules, even under conditions of high relative humidity, and in transferring the mixed sample into the die. The pressed pellets are durable and are stable. Pellets have been prepared with highly reactive compounds such as sodium perxenate, xenon trioxide, permanganate, and periodates by careful handling. In the case of nonreactive materials, spectra have been run on the same pellet after six weeks storage with no alteration in spectral characteristics of matrix transmittance.

The pressed silver chloride pellets may be handled with no special precaution except protection from direct sunlight. Some blank pellets have been used for several full days of infrared work in a fully lighted (fluorescent) room with no notable loss of transmittance.

The die faces must be clean and well polished as suggested by Metzler.[1] The occasional touching up with fine grinding compound and ammoniacal ethanol (1 cc conc. NH_4OH in 100 cc EtOH 95%) keeps the die in good condition.

Acknowledgement. The authors are indebted to Dr. Howard Claasen of Wheaton College for the original suggestion to attempt pelleting with silver chloride.

2.6 Pressed-Film Technique for Obtaining Infrared Spectra of Thermoplastic Materials*

J. E. Coakley and H. H. Berry

U. S. Naval Avionics Facility, Materials Laboratory and Consultants Division, Organic Materials Branch, Indianapolis, Indiana

Various techniques heretofore have been used in preparing cured thermoplastics for infrared analysis. Foremost among these techniques are: (1) dissolving in a solvent and casting of a film, (2) mulling, (3) pyrolysis, and (4) pressed disks. These techniques may be laborious, time-consuming, or impractical for materials that are difficult or impossible to dissolve, mull, or pelletize.

*The work herein described was performed by the authors at the U. S. Naval Avionics Facility, Indianapolis, Indiana.

This paper describes a simple, rapid, and inexpensive compression-molding technique. In most cases, films produced by this technique yield good spectra when properly prepared. Some pressed films were too thick to obtain good spectra of materials with very intense absorption bands. The celluloids, for example, contained some bands that were structureless because they were far too intense. This limitation does not impede identification when reference spectra are available.

This investigation covered all of the commercially important thermoplastics, including the newer ones. These materials were in the form of cured plastics such as sheets, rods, films, moldings, wire insulation, fabrics, and fibers.

Since it was not feasible to include pressed film spectra of each material, this report includes a few selected examples. In order to compare pressed-film spectra with those obtained by standard techniques, it was necessary to select some materials (in a suitable form) that are readily adapted to conventional infrared sampling.

Equipment used to prepare the film consisted of a mold (two flat, polished chrome–nickel stainless-steel disks, 3.5 cm in diameter × 2.4 cm high), a hot plate, and a laboratory press.

The following procedure, with some practice, produced thin transparent films that yielded good infrared spectra. Specimens approximately 1.5 cm square and 0.2 cm thick were cut from the samples. Such a specimen was laid on the polished side of one of the disks, and both disks were placed on a previously heated hot plate with their polished sides up. The specimen was probed at frequent intervals with

a stainless-steel spatula to determine when its soft-
ening point had been reached. At this point, the
polished side of the second disk was placed on top
of the specimen. Using tongs, and keeping the sides
of both disks even, the mold containing the specimen
was transferred to a laboratory press and pressed
with a force of approximately 15 000 lb. The film was
allowed to cool under pressure in the mold. The hard-
ening, or cure, time is dependent upon the material,[1]
preheat temperature, and thickness of molded film.
It may be a few seconds (for Teflon) or several minutes (for Kel-F and nylon). Pressure on the mold
was released, the disks were separated, and the film
was recovered. The use of highly polished chrome–
nickel stainless-steel disks for the mold and of a mold-
release agent such as a heat-resistant silicone fluid
will, in most cases, prevent the film from sticking to
the mold. Sometimes the upper and lower halves of
the mold remained stuck together and were separated
in the press. When the film remained stuck to one
half of the mold, it was removed with the aid of a
single-edge razor blade. It was not necessary to
remove the film intact. A piece of the film large
enough to cover the effective area of the spectrometer
beam produced a good spectrum.

Films prepared in this manner were mounted on
a solid-sample holder with plastic tape and placed
in the spectrometer beam for an infrared scan. The
principal criterion of quality was a clear film (0.005–
0.035 mm thick) showing no visible signs of decomposition and having the strongest absorption band in
its spectrum in the 2%–10% interval. Spectra were

1. In order to obtain clear films of nylon and Kel-F, slow
cooling in the press is necessary before making the infrared examination.

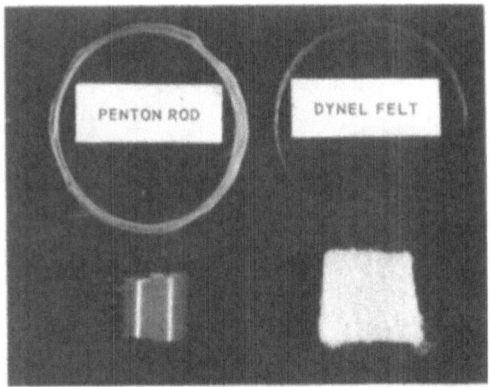

FIG. 1. (a) Penton. Lower left: original material; upper left: pressed film. (b) Dynel felt. Lower right: original material; upper right: pressed film.

obtained with a Perkin–Elmer model 21 double-beam spectrophotometer equipped with sodium chloride optics. Studies were limited to the 2.5-to-15.0-μ portion of the spectrum. No efforts were made to obtain quantitative data.

A photograph (Fig. 1) shows two original materials and their pressed films. Shown in Fig. 2 are spectra of an acrylic ester[2] as it appears in a pressed film, a Nujol mull, a potassium bromide disk, and a pyrolyzate. The pressed-film spectrum bears a close resemblance to that of the pressed disk. With the exception of bands at 3.45, 6.80, and 7.20 μ (interference from Nujol), the Nujol spectrum closely resembles those of the pressed film and pressed disk. The pyrolyzate spectrum, however, bears the least resemblance to that of the pressed film.

2. Acryloid KM-228, by Rohm & Haas. The use of commercial product names in this paper should not be interpreted as an endorsement by the U. S. Navy.

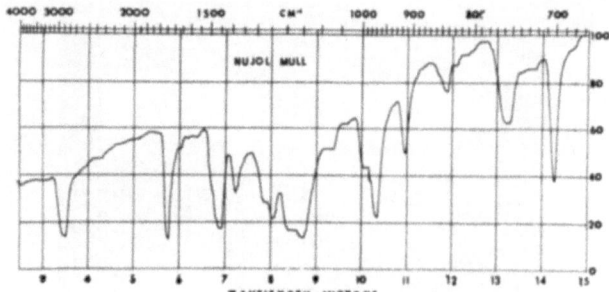

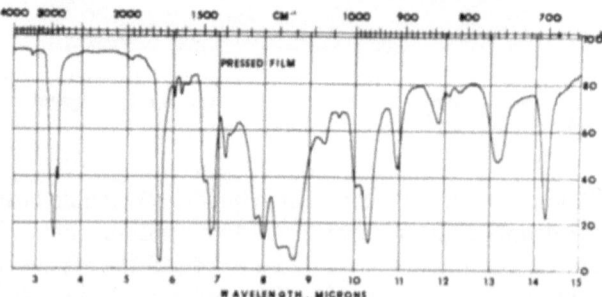

Fɪɢ. 2. Spectra of

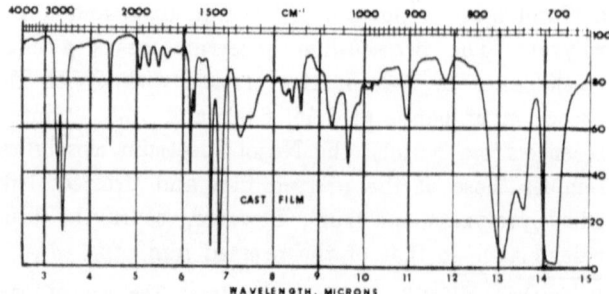

Fɪɢ. 3. Spectra of

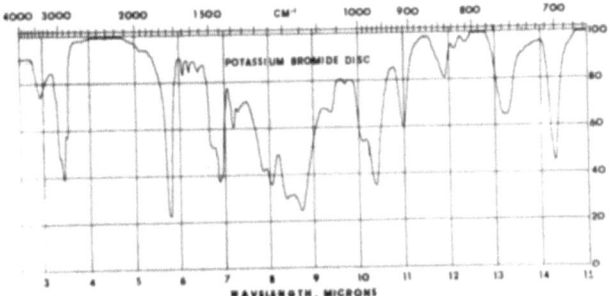

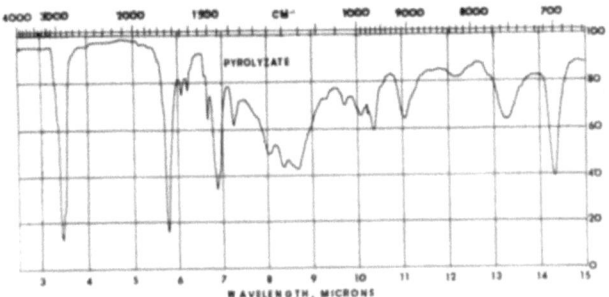

Acryloid KM-228.

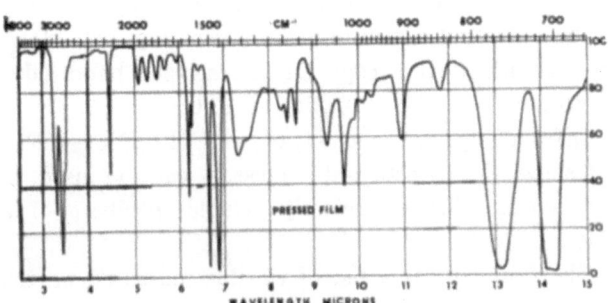

styrene–acrylonitrile copolymer.

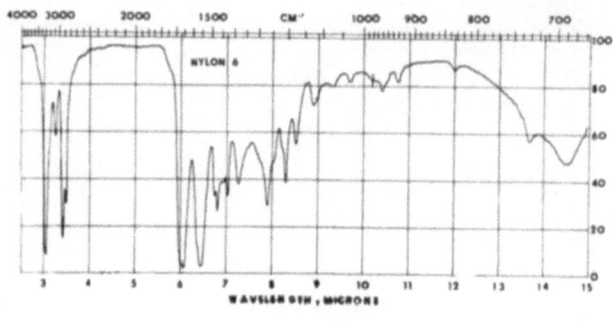

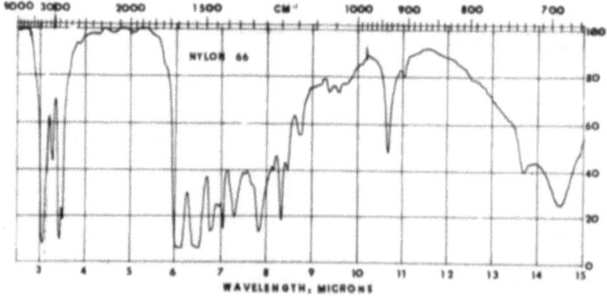

FIG. 4. Pressed-film

Presented in Fig. 3 are infrared spectra of styrene acrylonitrile copolymer (Tyril from Dow) obtained from a pressed film and from a cast film. The spectra are identical except for the solvent band (due to toluene) in the cast film at 13.63 μ.

Infrared spectra of 6, 11, 66, and 610 nylons (Fig. 4) show that pressed-film spectra can be used to differentiate between closely related members of a class.

The following is a list of conclusions based on the investigation described herein:

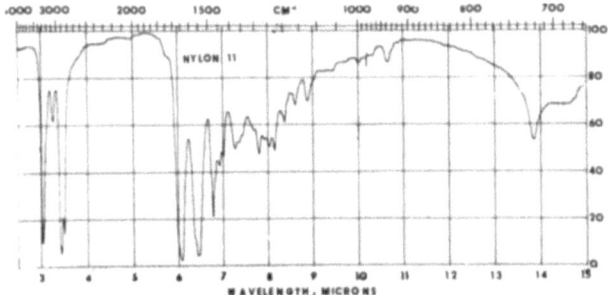

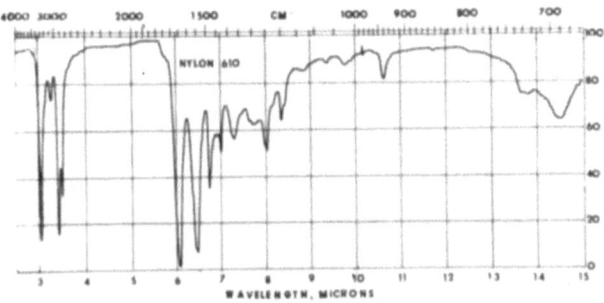

spectra of nylons.

1. The pressed-film technique provides a rapid and inexpensive method for preparing films of thermoplastics for infrared examination.

2. Films can be readily prepared by technicians with little or no additional training.

3. The ease of sample preparation afforded by this technique makes it preferable to conventional infrared-sampling methods for spectra of comparable quality.

4. Good pressed-film spectra can be obtained for thermoplastics that are insoluble, too resilient to mull,

or not readily adaptable to pyrolysis or pressed-disk techniques.

5. The pressed-film technique is suitable as a supplement to—or, in some cases, as a substitute for—other methods of infrared sampling.

The authors wish to thank Stanley Wlasik for assistance with the design and fabrication of the mold.

2.7   Pressed-Film Technique

Dorothy Camer, Marion Martin, and Ernestine Medeck

Johnson & Johnson Research Center, New Brunswick, N. J.

The technique of pressing films reported by Coakley and Berry[1] is similar to one used in our laboratory. In our case, a $1\frac{1}{4}$-in. x-ray die with separate, removable die faces (pellets) is used in an air-driven press in which only the top plate is heated. When the sample is placed directly on the die faces, it is difficult to pry the die faces apart after the pressure is removed. The sample film is frequently torn in the effort to get the die faces apart. To correct this situation, the sample is sandwiched between two pieces of Teflon film cut to the size of the die face. The sample-Teflon sandwich is placed between the die faces and pressure is applied. When the pressure is removed, the die faces separate immediately from the sample-Teflon sandwich. The Teflon film is then easily stripped from the sample without danger of tearing. The sample-Teflon sandwich can be re-inserted between the die faces, if a thinner sample film is desired.

1. J. E. Coakley and H. H. Berry, Appl. Spectry. **20**, 418, (1966).

Infrared Cell for Reactive Gases   2.8

Edward J. Bartoszek, Charles J. Mackley, and
David M. Gardner

*Research and Development, Pennsalt Chemicals Corporation,
King of Prussia, Pennsylvania 19406*

In the course of our work with reactive fluorine
compounds, we have had some difficulties in determining the infrared absorption spectra of many reactive gaseous compounds by using the common metal
cell made of stainless steel, monel, or nickel. For example, we have observed that with a metal cell it is
very important to pretreat the cell carefully with
fluorine or other appropriate gases; otherwise the
sample will decompose appreciably by reaction with
oxides or hydrates on the metal. Corrosion is always a
problem, and the proper assembly of a metal cell is
critical to provide gas-tight seals without excessive
strain on fragile cell windows. In addition, silver
chloride windows require special gaskets to prevent a
rapid etching of the window by electrolytic action
of the silver chloride in contact with the metal.

We have eliminated all these problems by use of a
cell made almost entirely of polytetrafluoroethylene
(PTFE), illustrated in Fig. 1. The cell body is machined from a PTFE tube (83 mm long; 38 mm o.d.;
22 mm i.d.) and has grooves at each end (see detail)
which function as concentric gas-tight ring seals
against the window. Because of the difficulty of maintaining a PTFE threaded joint under tension for
periods of time, the necessary compression on the cell
windows is provided by encasing the cell body in a
standard-size polyvinylchloride (PVC) pipe [108
mm long; 48 mm ($1\frac{7}{8}$ in) o.d.; 41 mm ($1\frac{5}{8}$ in.) i.d.].
The ends of the PVC pipe are tapped and fitted with
threaded plugs having a 22-mm-o.d. hole in the center

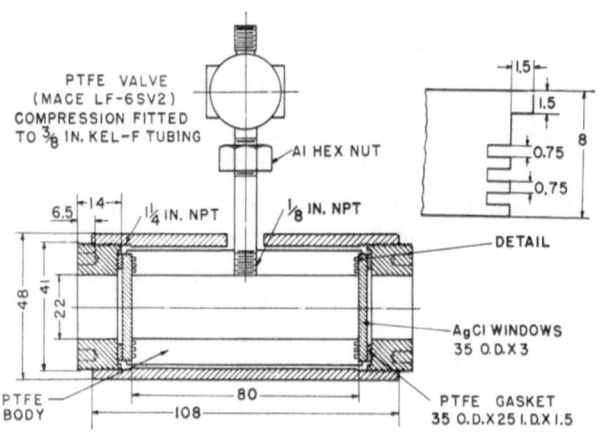

FIG. 1. Infrared gas cell (dimensions in millimeters).

to pass the infrared beam. These plugs are screwed tightly down onto PTFE gaskets to avoid scoring the windows. Both of the end plugs have three holes (see Fig. 1), 120° apart and 6.5 mm deep, to accommodate a face spanner wrench made specifically for tightening the plugs and forming the seal between the end of the cell and the window. Although the drawing indicates a silver chloride window, the seal works equally well with the more fragile calcium fluoride and barium fluoride windows.

As shown in the drawing, the gas sample is introduced through a short section of standard-size $\frac{3}{8}$-in. Kel–F tubing, which passes through a hole in the PVC case and is screwed directly into the PTFE cell body. Care must be taken in tapping the hole and threading the Kel–F tube, so that the end of the inlet tube is flush with the inner wall of the cell. This is most easily done by allowing the tap to run the equivalent of two turns beyond the inner wall surface during the tapping operation. A thread too deeply cut will cause the inlet tube to protrude within the cell and block out part of the beam. The all-PTFE valve used to close off the inlet tube (Mace Corp. catalog No. LF–6SV2) is compression fitted to the inlet tube and has given excellent service.

For some special studies, our cells have been equipped with a condensation well made of a similar piece of Kel–F tubing threaded in the same manner as above and mounted 180° around the cell from the sample inlet tube. The condensation well is thermally sealed prior to threading into the cell by heating the Kel–F over a low flame until it softens, and then squeezing tightly until cold in a small vise with smooth-faced jaws. It has been found that a second heating followed by a quench in cold water will make the lower end of the well nearly as transparent as glass.

Cells of this design have been used in studies of a wide variety of reactive gases, and they have distinct advantages over the comparable metal cells. No passivation or any type of pretreatment of the cell is required, and no substances other than the window material and inert PTFE and Kel–F come in contact with the sample. In addition, the compression seals used in the cell have proven to be very dependable, and gas leaks are rare even after extended periods of normal laboratory use.

High-Temperature Vacuum Cell for Infrared Studies of Adsorbed Molecules

2.9

R. O. Kagel and L. W. Herscher

The Dow Chemical Company, Chemical Physics Research Laboratory, Midland, Michigan 48640

A high-temperature all-metal cell suitable for infrared adsorption studies has been described in the liter-

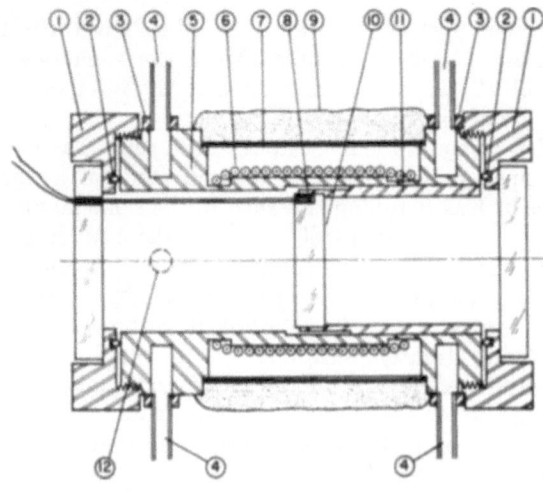

FɪG. 1. Adsorption cell: (1) window end cap; (2) Viton O-ring seals; (3) ring seal closures of water grooves; (4) cooling-water ports; (5) cell body; (6) heater winding; (7) aluminum heat reflector; (8) thermocouple; (9) asbestos insulation; (10) KBr-crystal sample support; (11) sample-plate-holder cylinder; (12) gas inlet, behind this point (not shown in this sectional view).

ature.[1] Our applications, however, require a higher operating temperature (>700°C), better vacuum (~ 10^{-6} Torr), and a larger aperture. The higher operating temperature requires a higher temperature gradient between the inside and outside, so that the surface of the furnace section does not become unreasonably warm. This temperature gradient results from a more efficient furnace design which utilizes reflection principles for confining the heat inside the furnace section. We have also found our method of construction and method of introducing the thermocouple

1. F. R. Harrison and J. J. Lawrence, J. Sci. Instr. 41, 693 (1964).

into the system more convenient for this type of study.

Figure 1 is a diagram of our cell, drawn to scale. The minimum aperture diameter for the light path is $1\frac{1}{2}$ in. The over-all length with window caps in place is $5\frac{1}{4}$ in. The cell body is essentially a hollow cylinder of stainless steel. Marsh beaded heater wire is used for the furnace winding. This material is No. 19 gauge Nichrome wire threaded through short sections of ceramic tubes or beads and requires no additional insulation. The thin-walled sections of the cell body at either end of the heated section restricts the flow of heat to the cooled end-caps. There is a $\frac{3}{16}$-in. airspace insulation between the bright aluminum reflector shield and the heater winding. For additional insulation, the reflector shield is covered with a thick layer of asbestos.

When the sample region is heated to 720°C, the outer surface of the asbestos reaches about 375°C. The power required to operate the sample region at 720°C is 820 W (105 V at 7.8 A). The temperature limit appears to be determined by the melting point of the heater winding. This limit has not yet been determined, although the maximum temperature attained thus far was about 775°C.

The brass end caps to which potassium bromide windows are sealed are maintained at room temperature to avoid thermal strains in the crystals. Most gaseous samples are studied at sufficiently low pressures so that condensation is not a problem. However, when condensation does become a problem, the windows can be heated up to 50°C and the system evacuated. The window temperature is controlled by the rate of water flow through the passages provided at either end of the main cell body. These passages are formed by machined annular grooves in the cell body which are closed by stainless-steel rings. The rings include inlet and outlet port tubes of stainless steel.

The ring and port assemblies are welded to cover the grooves with a water-tight seal by the spray-fuse silver soldering process.

The potassium bromide windows are sealed to the end caps with D.E.R.® 331 epoxy resin. The end caps are assembled onto the cell body by a screw thread and vacuum sealed with O rings made of Viton. This arrangement allows for easy disassemble and access to the sample-supporting plate, which is held by a removable stainless-steel cylinder. The sample-supporting end of the cylinder is slotted at regular intervals in order to grip the sample plate and to facilitate evacuation. The sample is reproducibly positioned by means of a shoulder stop at the center of the cell body.

Thermocouple leads for measuring the sample-plate temperature are brought in through a hole in one of the potassium bromide windows. The leads are vacuum sealed with epoxy resin.

The cell is currently being used to investigate the adsorption of various alcohols on oxide surfaces.[2]

2.10   Sample-Inlet System for Studying Adsorption of Low Vapor Pressure Materials on Solid Surfaces

R. O. Kagel

Chemical Physics Research Laboratory,
The Dow Chemical Company, Midland, Michigan 48640

Infrared spectroscopy has not been extensively used to study vapor-phase adsorption of low volatility liquids or solids on solid surfaces. This is almost

2. R. O. Kagel, J. Phys. Chem. 71, 844 (1967).

certainly because of the difficulties involved in introducing these samples into the cell in such a way that
they can interact with an activated solid (catalytic)
surface. In this type of application, it is necessary to
devise a sample-inlet system which has a small volume,
is operable at temperatures up to 200°C, is evacuable
to 10^{-5} Torr, and will not interfere with the high
temperature (500°C) oxidation (1 atm O_2) or high
vacuum (10^{-5}–10^{-6} Torr) pretreatment of the catalytic surface. That is to say, the inlet system must be
an extension of the adsorption cell itself.

This note describes a sample-inlet system which
satisfies the above requirements. The design offers
several additional advantages: (1) liquid samples may
be introduced directly onto the solid surface (liquid–
solid interaction studies); (2) vapors may be flushed
into the cell in the immediate vicinity of the solid
surface without condensation (vapor–solid interaction studies); (3) the infrared beam is not appreciably blocked; and (4) the inlet system is easily
detached from the cell, dismantled, and assembled.

The sample-inlet system (Fig. 1) was designed to
fit the adsorption cell described by Kagel and
Herscher.[1] The system consists of $\frac{1}{8}$-in. stainless steel
tubing A. The valves are welded bellows valves with
the exception of V-1 which is a welded bellows-needle
valve. V-1 acts as a crude metering valve. The portion
of the line extending into the cell B is $\frac{1}{16}$-in. stainless
steel tubing; this probe ends a few millimeters from
the solid surface D. The bottom half of the stainless
steel sample holder C is removable; a Viton O-ring
provides a vacuum-tight seal between the two halves
of C. V-3 connects the inlet system to a furnace F and
an inert-gas (N_2) cylinder. The entire inlet system is
wrapped with heating tape. It is compact and when
attached to the adsorption cell, conveniently fits into

1. R. O. Kagel and L. W. Herscher, Appl. Spectry. 21, 187
   (1967).

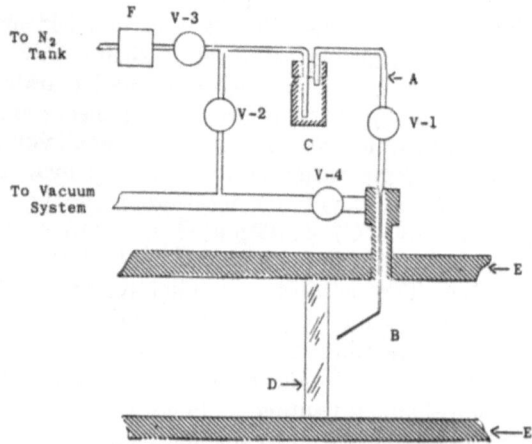

FIG. 1. A, $\frac{1}{8}$-in. stainless steel tubing; B, $\frac{1}{16}$-in. stainless steel tubing; C, sample container; D, solid sample; E, adsorption cell body; F, furnace; V-1, bellows-needle valve; V-2, V-3, and V-4, bellows valves.

the sample compartment of a Beckman IR-9 spectrometer.

Sample is placed in container C. C is attached to the inlet system, cooled (dry ice or liquid N₂), and the system is then evacuated to ∼10^{-5} Torr. V-2 is closed while the sample is allowed to warm up to room temperature. The entire inlet system may be heated up to about 200°C. (The solid surface temperature is independently controlled.) Liquid can be pumped directly into the cell through V-1 (V-2 and V-3 closed), or vapors can be flushed into the cell with a stream of hot (150°–200°C) inert gas (V-3 open). Vapors which have been carried into the cell adsorb on the solid surface while excess carrier gas is removed by evacuation. The amount of sample introduced into the cell is controlled by the flow rate of inert gas (when V-3 is open). Experience has shown that the optimum flow rate is such that complete surface coverage is achieved in about two hours.

Method for Rapid Transfer of GLC   2.11
Fractions into Infrared Cavity Cells

R. F. Kendall

*Bartlesville Petroleum Research Center, Bureau of Mines,
U. S. Department of the Interior, Bartlesville, Oklahoma*

Combining gas–liquid chromatography (GLC) and infrared spectrophotometry is an established analytical practice. Special trapping procedures are required, however, for the collection and successful infrared analysis of microliter samples obtained from GLC instruments. Procedures designed for the collection of volatile GLC effluents include: (1) Trapping in powdered potassium bromide, followed by pelleting; (2) trapping in thermoelectrically cooled cells, for use with multiple reflection attachments; (3) trapping in capillary tubes containing solvent; and (4) trapping through a special fraction collector into ultramicrocavity cells, which are used with beam-condensing equipment.

Experience has shown that the cavity-cell technique is effective in analyzing volatile petroleum fractions.[1,2] Further, the sample is recoverable from the cavity cell for analysis by mass spectrometry or by other types of instrumentation, and solvents are not required for collecting GLC fractions in quantities greater than 0.6 μl.

Commercially available units employ a fraction-collecting device which is used to trap the GLC effluent by condensation directly into the cavity cell. This is accomplished by covering the cell with a metal

1. H. J. Coleman, C. J. Thompson, R. L. Hopkins, and H. T. Rall, J. Chem. Eng. Data **10**, 80–84 (1965).
2. C. J. Thompson, H. J. Coleman, R. L. Hopkins, and H. T. Rall, J. Chem. Eng. Data **9**, 473–479 (1964).

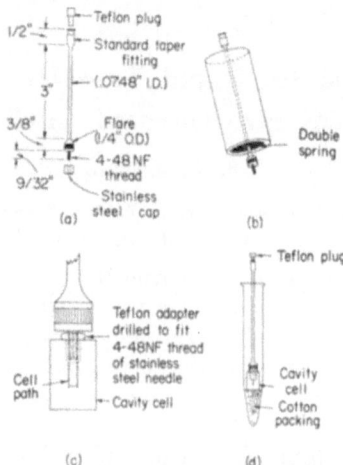

Fig. 1. Equipment for transferring sample from chromatograph to infrared cell: (a) stainless-steel trap, (b) cooling container, (c) Teflon adapter, and (d) apparatus contained in centrifuge tube.

cap and immersing the unit in a cooling solution for 15 min before trapping a GLC fraction. After sample collection, the unit is warmed to ambient temperature before removing the metal cap, in order to minimize moisture condensation on the outer surface of the NaCl cavity cell, thus requiring an additional 15 min.

An improved trapping and transfer procedure has been developed which eliminates the two major disadvantages mentioned, namely, excessive time required and moisture condensation on the infrared cell. A trap made of stainless-steel needle stock [see Fig. 1(a)] is used for collecting the GLC fraction, and the trapped material is then centrifuged directly into the cavity cell. Only 3 min are required for transferring the sample. Another advantage is that several samples can be collected in individual traps from a single chromatographic charge and stored, several hours if necessary, for subsequent infrared analysis.

Before collecting a GLC fraction, the trap [shown in Fig. 1(a), and previously described[3]] is covered with a stainless-steel cap and Teflon[4] plug and placed in a cork-lined brass cooling container, as shown in Fig. 1(b). A double spring prevents the trap from slipping through the container. The container is packed with finely powdered dry ice approximately 2 min before sample collection.

When the recording potentiometer indicates sample elution, the cap and plug are removed from the trap, and the trap is connected to the heated exit of the chromatograph. After the sample is collected, the container and trap are removed from the GLC exit, and the stainless-steel cap and Teflon plug are replaced to prevent moisture condensation of the inner surface. The material can now be transferred into the cavity cell or stored at dry-ice temperatures for subsequent analysis.

In loading the cavity cell, the trap is lifted from the container, and the stainless-steel cap is removed. The threaded portion is wiped free of moisture and then fitted with a Teflon adapter [Fig. 1(c)] made by drilling a standard Barnes Engineering Co. type D cavity-cell Teflon stopper. The remainder of the trap is wiped free of moisture and then connected to the cavity cell.[5] The entire unit is placed into a bench-type laboratory centrifuge tube as shown in Fig. 1(d). Cotton packing prevents the cell from

3. C. J. Thompson, H. J. Coleman, C. C. Ward, and H. T. Rall, Anal. Chem. **32**, 424–430 (1960).
4. Reference to specific brand names is made to facilitate understanding and does not imply endorsement of such brands by the Bureau of Mines.
5. The Teflon adapter must fit snugly but not tightly, otherwise the cavity cell will fracture during centrifugation.

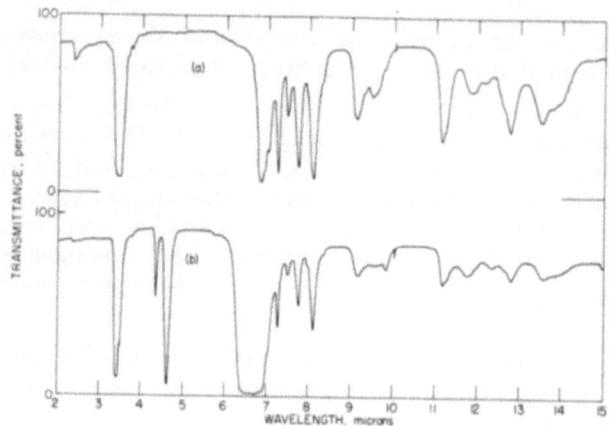

FIG. 2. (a) Infrared spectrum of 4-thiaheptane
collected from a GLC charge of 0.6 μl. (b) Infrared spectrum of 4-thiaheptane collected from
a GLC charge of 0.1 μl using 2.5 μl carbon disulfide solvent.

lodging in the tapered portion of the tube. The apparatus is centrifuged for 2 min at approximately
3400 rpm.

Solvents are not required for samples in quantities
greater than 0.6 μl if the boiling point of the material
exceeds 100°C. For smaller samples, and those volatile
at ambient temperatures, 2.5 μl of solvent are added
to the trap immediately before centrifugation.

Examples of the spectra obtainable with this trapping procedure are shown in Fig. 2. Curve (a) in
Fig. 2 is the spectrum of the compound 4-thiaheptane
collected from a GLC charge of 0.6 μl. Curve (b) in
Fig. 2 is the spectrum of a 0.1 μl charge of 4-thia-
heptane collected using 2.5 μl of carbon disulfide
solvent. Although solvent absorption bands are apparent at 4.4, 4.6, 6.3–6.9, and 11.7 μ, the remaining
bands of 4-thiaheptane are adequate for qualitative

interpretation.[6] Identifiable spectra have been obtained for GLC charge quantities as small as 0.05 μl by using solvent and 5× ordinate-scale expansion.

The spectra presented were recorded on a Perkin–Elmer model 21 spectrophotometer equipped with a 4× beam condenser. Type D cavity cells of 0.05-mm path were used for these experiments.

Simple Device for Preparing Polyethylene Liquid Cells for the Far Infrared

2.12

John Lyford, IV, and Walter F. Edgell

Department of Chemistry, Purdue University, Lafayette, Indiana 47907

Since it is difficult to remove solutions from the thin polyethylene cells which are frequently used in far infrared spectroscopy, it often becomes necessary to use a new cell for each measurement. Consequently, it seems desirable to describe a device which is suitable for fabricating these cells rapidly and inexpensively. The resulting cells have two further advantages for our work, much of which relates to oxygen and moisture sensitive solutions. Air tight seals are readily obtained with these cells, and the cells can be made of varying thicknesses of polyethylene. These last two properties are in contrast to those of the commercially available cells, which are sealed only with difficulty and which are available in only one thickness of polyethylene.

6. The absorption band at ~9.8 μ [(b) of Fig. 2] is from the silicone oil used as the GLC-column liquid phase. This material was found to exude from the column at elevated temperatures and, in this instance, was trapped with the compound 4-thiaheptane.

FIG. 1. Cell-making device.

The apparatus is designed to hold a sandwich consisting of two pieces of polyethylene with a spacer in the middle while heat is applied to cause the polyethylene to seal. After the cell has been formed and the spacer removed, the cell is filled through a syringe needle. It is then sealed by heating and crimping the open end. The apparatus is quite adaptable to the use of various thicknesses of polyethylene for the cell preparation, and the resulting cells are readily filled and sealed. Pure liquids are quite easily removed from the cells, but with certain solutions there is some difficulty in completely cleaning the cell. However, in these cases no problems are encountered in obtaining the spectrum of both the solution and its pure solvent in the same cell if the solvent is run first.

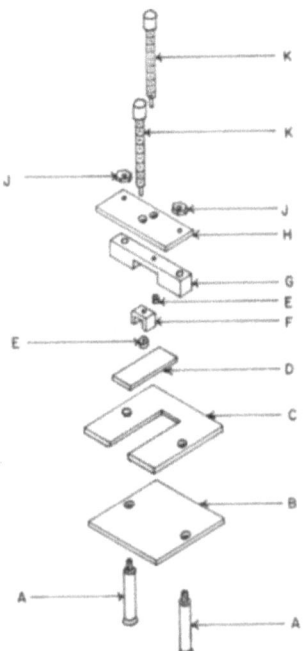

FIG. 2. Exploded view of apparatus.

A photograph of the apparatus is shown in Fig. 1, and an exploded view of the device is given in Fig. 2. Only a few of its dimensions need attention. These are the size of the shoe D, the distance between the guideposts A, and the size of the slot in part C. The distance between the guideposts should be as great as the width of the cell holder in the instrument, while the dimensions of the shoe should be at least as large as those of the opening in the cell holder through which the beam passes.

The slot which is cut into part C should have the same dimensions as the shoe, which fits into it. More-

over, careful examination of Fig. 2 will reveal that
the lower edge of this slot is beveled. Since the bevel
is required to prevent part C and the spacer from
exerting a scissors effect on the polyethylene, the
amount of the bevel is dictated by the desired thickness of the cell as well as by the thickness of the
polyethylene itself.

The device is constructed of brass. The two screws
K are not threaded below the point at which they
pass through the connecting pieces F and G, and
the diameter of the rod is decreased by $\frac{1}{2}$ below this
point. These screws are attached to their respective
connecting pieces F and G by passing the smooth rod
through the hole in the connecting piece, slipping the
brass rings E over the ends of the rods, and soldering
the rings in place. Parts F and G are then soldered
to parts D and C, respectively. Since the holes in
plate H through which the screws K pass are
threaded, the position of parts C and D can be easily
adjusted with these screws.

The diameter of the guideposts A is decreased by
$\frac{1}{2}$ above the point at which they pass through plate
H. Since they are also threaded above this point, the
plate is conveniently held in place by the two nuts J.

In the preparation of a cell, a piece of metal somewhat smaller than the shoe is used as a spacer. The
thickness of the finished cell is determined by the
thickness of this spacer. It is sandwiched between two
pieces of polyethylene, at least one of which should
have the proper dimensions to fit the cell holder of
the instrument which is to be used. After this sandwich has been placed between pieces B and C of the
apparatus, the two screws K are tightened until the
sandwich is firmly held in place. The entire assembly
is then heated in a conventional drying over to allow
the polyethylene to seal. The length of heating time
is important. If it is too short, a good seal is not

made. On the other hand, if it is too long, the polyethylene tends to flow and to stick to the spacer. We have found that for polyethylene which is 19 mils thick the optimum time is 31–33 min at a temperature of about 145°C. If thicker polyethylene is used, slightly longer heating times are required.

.After the cell has been formed and the spacer removed, a syringe needle is inserted between the pieces of polyethylene. The open end of the cell is then heated with a hot soldering iron, and the edges are crimped. When the cell has been filled through this needle, it is withdrawn; and the resulting hole is sealed by heating and crimping the edge at this point. After the spectrum has been obtained, the cell can be cleaned by cutting off the top and inserting a syringe needle. After cleaning, the cell may be resealed and used again in many cases.

We have made cells by this process over the thickness range 0.025–2.0 mm. There seems to be nothing, however, which would prevent thicker cells from being made as long as the slot in part C is provided with sufficient bevel. Thinner cells could be made if spacer material of sufficient strength could be found. The thickness of the polyethylene which was employed depended upon the nature of the sample. For some liquids we found that a thickness of 19 mils was sufficiently rigid to provide a firm cell, whereas for others we were obliged to use polyethylene in thicknesses up to 50 mils.

We would like to thank Dr. Vernon Thornton of the Phillips Petroleum Company, Bartlesville, Oklahoma, for a generous gift of their Marlex brand of polyethylene. We are also grateful to John V. Hession for fabricating this device, and one of us (J. L.) is indebted to the Applied Research Projects Agency and to the Minesota Mining and Manufacturing Company for financial support.

2.13   Method for Making Disposable Far-Infrared Cells with Fixed and Variable Pathlengths

E. Kinsella and J. Coward

Chemistry Department, The City University,
London, England

The manufacture and applications of three types of polythene liquid cells, for use in far-infrared spectroscopy, are described. The cells are very cheap and simple to make, and may reasonably be considered as disposable. They are particularly useful for the examination of air-sensitive liquids. No elaborate equipment is required for the manufacture of these cells, though the method outlined below is capable of considerable refinement should this be thought necessary in a particular case. The three types we have used are (1) nominally-fixed pathlength cells, (2) multiple-fixed pathlength cells, (3) variable pathlength cells. The cells described below are designed for use with a Grubb-Parsons DM4 spectrometer so dimensions might have to be altered for other instruments.

The cells are moulded from rigid high-density polythene sheet of thickness 0.030 in.–0.040 in. The sheet is guillotined into rectangular pieces 3 in.×2 in. for multiple- and variable-pathlength cells, and 2 in. ×1½ in. for fixed-pathlength cells. The cells are pressed between two sheets of springy metal (5 in.×3 in.) hinged along their narrow edge, such that, when laid open they lie absolutely flat, and, when closed parallel to each other, the gap between them is 0.080 in. The two plates are cut preferably from 0.020-in. thick, polished stainless steel, but heavy gauge tinplate is satisfactory if care is taken not to give it a permanent bend.

The hinged plates are placed open on a hotplate set at the temperature of melting polythene. Two of the pieces of polythene are placed on the plates such that when the plates are closed the polythene pieces will lie on top of each other. The polythene is pressed onto the plates until it melts and the appropriate former (see Fig. 1) laid on one of the pieces of polythene. The former is laid lengthwise along the polythene, by placing one end (the thin end in the variable and multipathlength cells) $\frac{1}{4}$ in. from the end of the polythene nearest the hinge, and then pressing down flat, allowing time to squeeze out any air bubbles. The other half of the hinged plate is then folded over and pressed onto the first plate, again, in such a way as to remove air bubbles. Firm pressure is now applied until the polythene conforms to the former and produces a good seal around it. The time required for this depends on the thickness of the polythene and

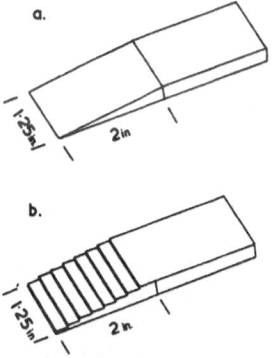

FIG. 1. (a) Former, for variable-pathlength cell, machined, as shown, from 4 in. ×1¼ in. ×(0.080 in. or 0.040 in.) spring or stainless steel. Dimensions may be altered to suit requirements. (b) Former, for multiple fixed-pathlength cells, machined, as shown, from the same material as for (a). The step width is ¼ in. and the step height is 0.010 in. for 0.080-in. thick steel, and 0.005 in. for 0.040-in. thick steel.

former, and has to be established by trial and error. In the case of the variable-pathlength cells, the pressing is carried out so as to give a taper along the cell, so that the cell-wall thickness is approximately constant.

The assembly is then removed from the hotplate and cooled while still under pressure. The cell is removed from between the two plates (the springiness of the steel facilitates this step) and the former withdrawn after flexing the cell to free it. Two wires, of the same gauge as the hypodermic needle used for filling the cell, are inserted in the open end of the cell and separated by a strip of polythene to a depth of $\frac{3}{4}$ in. This end of the cell is sandwiched between the two heated plates until it melts. Pressure is applied to ensure a good seal and the assembly is removed and cooled as before. After removing the cell from between the two plates, the wires are withdrawn to leave two capillaries leading into the cell compartment. If required, the edges of the cell may be trimmed off in a guillotine so that it will fit an appropriate cell holder.

With fixed-pathlength cells of 0.010 in. pathlength or less, a modified procedure is adopted to facilitate the filling of the cell. The polythene pieces are placed on the hinged plates as before and the former, cut from a suitable length of shimming strip, is laid on one of them. A wire, bent as shown in Fig. 2, is then carefully placed on the polythene so that its ends just overlap the former. The other half of the hinged plate is then folded over and the cell formed, cooled, and removed from the assembly as before. The former is then withdrawn and the top $\frac{3}{4}$ in. of the cell pressed between the hot plates to seal it. The wire is withdrawn after the cell has cooled.

The above operations can be carried out using separate metal or glass plates instead of the hinged assembly, but more care is required. When glass plates are being used several difficulties may arise; the polythene is often very hard to remove from the glass so

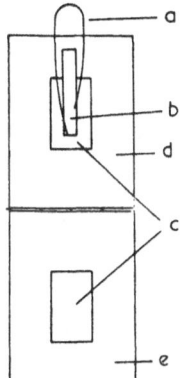

Fig. 2. Layout of components prior to pressing cell (0.010-in. or less fixed-path-length cells illustrated in this case): a—wire; b—metal former; c—Polythene; d, e—Hinged plates. Plate e is folded over onto plate d before pressing.

the plates have to be smeared with Silicone grease, which must then be washed off the completed cell. The glass also tends to crack unless borosilicate glass is used. However, this is fairly costly and has the disadvantage of taking a long time to heat up and cool down. The advantage of glass is that it is possible to see when all the air-bubbles have been squeezed out of the cell-compartment area.

Difficulty may also be encountered in the withdrawal of the former. To overcome this, it is advisable to have a high polish on the metal and to give it a slight taper across its width. The edges, in particular, should be rounded and smooth. Silicone grease may be used, but this must be flushed out of the cell before the final sealing.

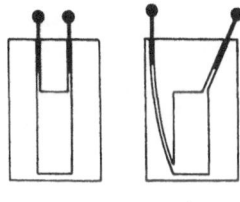

Fig. 3. Examples of fixed pathlength cells. (a) Cells >0.010-in. pathlength. (b) Cells <0.010-in. pathlength.

All the cells described are particularly useful for air-sensitive materials since they can be filled from a syringe in a drybox and the long capillary lead-in minimizes the effect of any leakage after stoppering with platinum wires. A further advantage is that a large number of cells can be filled in the drybox and taken out together, thus obviating the time-consuming process of transferring a conventional liquid cell through the entry system several times. Samples may be left in the cells for a reasonable length of time though with many organic solvents the polythene begins to distort appreciably after a few hours. In many cases, the cells can be flushed out and used repeatedly, but this is usually difficult with those of short pathlength. The multipathlength cell is particularly suitable for running a spectrum of a liquid for the first time. It avoids the filling and running of a large number of cells of different pathlength. Having established the required pathlength, it is more economical to use the fixed-pathlength cells owing to their lower volume. The variable-pathlength cell is useful as a reference cell to obtain exact balancing of solvent absorption, as this is otherwise hard to achieve with polythene cells owing to the tendency of the polythene to distort. Where no importance is attached to the value of the pathlength, the variable cell is ideal for the sample as well as the reference, if sufficient material is available to fill it.

Cell Design for the Measurement of 2.14
Transmission Spectra of Liquids at Very
Long Infrared Wavelengths

G. C. Hayward

Beckman–R.I.I.C. Ltd., Worsley Bridge Road,
London, S.E. 26, England

Measurements of the infrared transmission spectra
of solids and liquids at moderate to high resolution
are often troubled by undesirable interference effects
caused by multiple internal reflection in parallel-
sided materials (commonly known as "channel spectra"), which usually become increasingly pronounced
at longer wavelengths. The purpose of this communication is to describe how a commercial vacuum-
tight liquid cell[1] may be modified to minimize such
interference effects arising both in the cell windows
and in the liquid cavity, and to present liquid transmission spectra recorded in the range 7–70 cm^{-1}
using such a cell in conjunction with a small commercial lamellar grating interferometer.[2,3] Briefly,
the amplitude of "channel spectra" observed in the
transmission spectrum of a material layer depends[4,5]
directly upon the transparency of the medium, the
effective reflection co-efficient at the material boundaries, and the degree of parallelism of the layer

1. FS-03 liquid cell manufactured by Beckman–R.I.I.C. Ltd.,
   Worsley Bridge Road, London, S.E.26, England.
2. LR-100 Lamellar grating interferometer manufactured by
   Beckman–R.I.I.C. Ltd., Worsley Bridge Road, London,
   S.E.26, England.
3. R. C. Milward, Infrared Phys. 9, 59 (1969).
4. See for example A. E. Martin, *Infrared Instrumentation
   and Techniques* (Elsevier Publ. Co., Inc., New York, 1966),
   pp. 122–126, Ref. 8, Sec. V1, 2.3.
5. C. R. Randall and R. D. Rawcliffe, Appl. Opt. 6, 1889
   (1967).

thickness compared to the ir wavelength of observation.

The "periodicity" of the interference fringes $\Delta\nu$ (cm^{-1}) is related to the material refractive index n, and thickness d by

$$\Delta\nu = 1/(2nd \cos \phi),$$

where ϕ is the average angle of beam incidence. The problem of channel spectra becomes noticeable when the instrumental resolution width is less than $\Delta\nu$. Interference in transparent liquid cell windows can be avoided if the cell windows are of wedge shaped section, a wedge angle of 3° being sufficient. A pair of bevelled windows can be fitted to a standard liquid cell by arranging them with the wedges opposed, as shown in Fig. 1.

It is possible to eliminate interference effects arising in the cell cavity by using a wedge-shaped cell spacer. This practice however, leads to an uncertainty

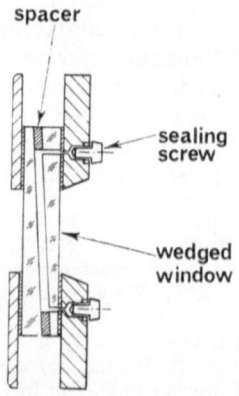

FIG. 1. Sectional view of liquid cell fitted with wedged windows. For clarity, wedge angle has been exaggerated.

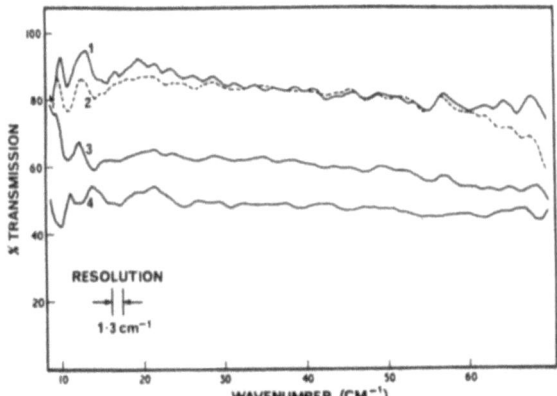

FIG. 2. Transmission spectra of window materials for use at very low frequencies, recorded at 1.3 cm⁻¹ resolution. (Thickness is in mm.) (1) Polypropylene (4 mm); (2) High-density polyethylene (3 mm); (3) Z-cut crystal quartz (3 mm); (4) High-purity silicon (2 mm).

in the cell path length, and moreover, is only practicable for cells of at least a few mm spacing, since a 3° wedge angle results in a difference of cell thickness of 1 mm over a typical cell aperture of 20 mm. Thus it is desirable to try and minimize cavity interference effects by matching the refractive index of the window material with that of the liquid to be studied, so that the reflection co-efficient at the liquid–window interface is minimal.

The transmission spectra of materials suitable for liquid cell windows for the wavelength region 7–70 cm⁻¹ are shown in Fig. 2. Of these, high purity silicon (resistivity >10 Ω cm for *n*-type material) has favorable chemical and physical properties, but its high refractive index (3.42 at 50 cm⁻¹)[5] results in a large reflection loss and severe interference within the cell cavity. It was therefore essential to wedge both cell cavity and windows when using silicon as

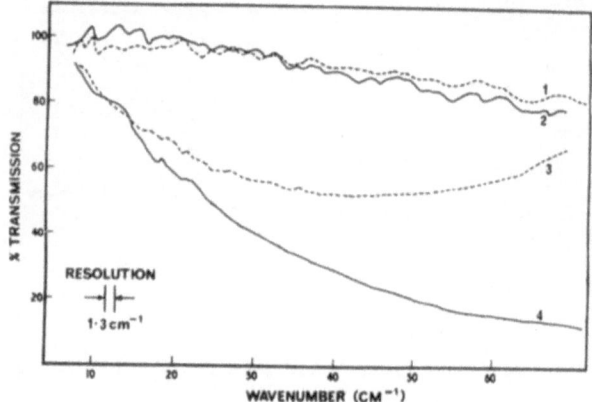

FIG. 3. Transmission spectra of liquids at 4-mm path length recorded at 1.3 cm⁻¹ resolution. (1) Cyclohexane; (2) *n*-Hexane; (3) Carbon tetrachloride: (4) Benzene.

window material. Crystal quartz has relatively low refractive indices (2.11 and 2.16 for the ordinary and extra ordinary rays, respectively at 50 cm⁻¹),[6] but its transparency range restricts its use to the lowest frequencies only. Polyethylene and polypropylene are inexpensive and have refractive indices close to those of many organic liquids.[7] Using liquid cells with the latter window materials and unwedged spacers, transmission spectra of a range of liquids were obtained which did not exhibit appreciable fringe structure, as shown in Figs. 3 and 4. In order to obtain a fringe-free "reference" spectrum of the empty cell however, it was essential to use a wedged spacer, due to the disparity in refractive indices.

For vacuum-tight sealing of cells with polymer windows it was necessary to use amalgamated lead

6. E. E. Russell and E. E. Bell, J. Opt. Soc. Am. **57**, 341 (1967).
7. G. W. Chantry, H. A. Gebbie, B. Lassier, and G. Wyllie, Nature **214**, 163 (1967).

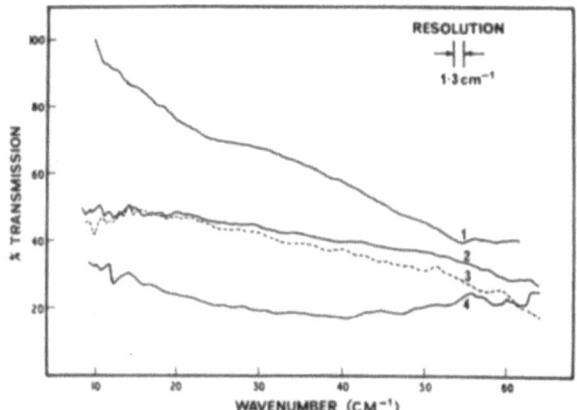

FIG. 4. Transmission spectra of liquids at various path lengths recorded at 1.3 cm⁻¹ resolution. (1) 1,4-dioxane (1 mm); (2) Water (nominal path-length 50 μm); (3) Acetonitrile (nominal path length 50 μm); (4) Chloroform (1 mm).

gaskets, while the other materials sealed effectively with PTFE (polytetrafluoroethylene) or amalgamated lead gaskets. It should be mentioned that if easily compressible gaskets or window materials are used, the pathlength of the cell will become uncertain, particularly for very short pathlengths. It may, moreover, become impossible to determine the pathlength by the interference method over the region of transparency of the window material; a cell of 50μ spacing, for example, gives fringes 100 cm⁻¹ wide.

Previous reports of the spectra of the liquids shown in Figs. 3 and 4 have included more restricted wavelength ranges.[7,8] The liquids studied were of Reagent grade purity. Except for the most transparent liquids, fringe patterns were eliminated, and the small oscillations at the edges of the spectral range resulted

8. A. Hadni, *Essentials of Modern Physics Applied to the Study of the Infrared* (Pergamon Press, Inc., New York, 1967), p. 468.

from transformed noise. The spectra show the broad absorption bands which are found in the spectra of many liquids at frequencies less than 100 cm⁻¹ and which are associated with restricted rotational motion in the liquid phase.[9]

In conclusion, polyethylene and polypropylene, by virtue of their high transparency and low refractive indices, appear to be the most suitable window materials for liquid cells for the far-ir region, 7–70 cm⁻¹. Their main disadvantage is that they are liable to become permanently contaminated by contact with organic liquids.

The transmission spectra of another polymer TPX (Imperial Chemical Industries), which is suitable for use as a far-ir cell window material and which does not appear to suffer from this disadvantage, has been reported recently.[10] These materials cannot be used in high temperature applications, and high purity silicon or crystal quartz may be used in such cases.

ACKNOWLEDGMENT

The author wishes to thank Dr. R. C. Milward for helpful advice in preparing this note.

9. J. E. Chamberlain, H. A. Gebbie, G. W. F. Pardoe, and M. Davies, Trans. Faraday Soc. **64**, 847 (1968).
10. G. W. Chantry, H. M. Evans, J. W. Fleming, and H. A. Gebbie, Infrared Phys. **9**, 31 (1969).

Far-Infrared Transmission of Commercially Available Crystals and High-Density Polyethylene* 2.15

W. G. Fateley, R. E. Witkowski, and G. L. Carlson

Mellon Institute, 4400 Fifth Avenue, Pittsburgh, Pennsylvania 15213

Recently, far-infrared spectrometers and interferometers have become available to many researchers and laboratories. With this new interest in the low-energy region of the spectrum, we thought it useful

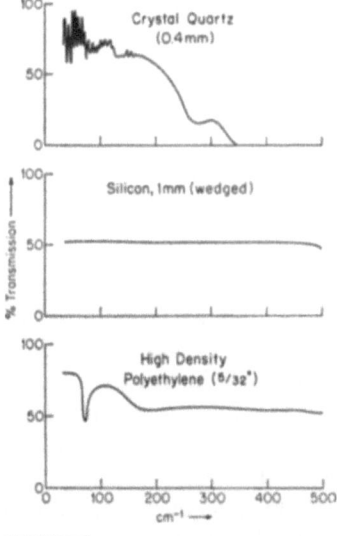

FIG. 1. The transmission curves of (a) crystal quartz cut so that the faces are parallel to the optical axis, 0.4 mm thick, (b) silicon with a 50-min wedge between the faces, 1 mm at largest thickness, (c) high-density polyethylene, $\frac{5}{32}$ in. thick.

*This research was supported in part by the National Institutes of Health, U. S. Department of Health, Education, and Welfare, Grant No. GM-11815, and the U. S. Air Force Command, Wright–Patterson Air Force Base, Ohio, under contract No. AF 33(657)11142.

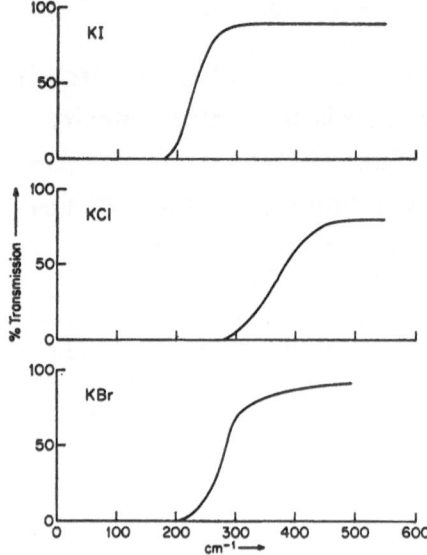

Fig. 2. Transmission spectra of KI, KCl, and KBr, 1 mm thick.

to publish a collection of transmission curves of commercially available crystals and high-density polyethylene so that choices could be made for cell windows and filters for work in this region.[1]

The spectra given in Figs. 1–6 were measured on the Beckman IR-11 grating far-infrared spectrometer and the RIIC model FS-520 far-infrared interferometer. Comparison of the spectra obtained on both of these instruments was very satisfactory. All crystals, unless otherwise indicated, were approximately 1 mm thick. The silicon was cut with a 50-min wedge angle between faces to eliminate fringing from internal reflections of the parallel surfaces.

1. We would like to thank Harshaw Chemical Co., Cleveland, Ohio, for providing us with the polished crystals.

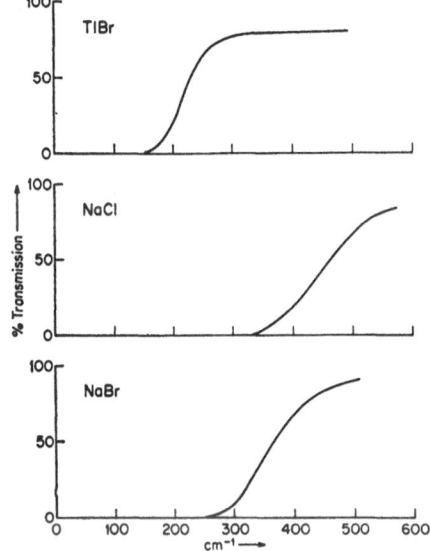

FIG. 3. Transmission spectra of TlBr, NaCl, and NaBr, 1 mm thick.

Many of the transmission curves have been reported by previous workers. In some of the cases, earlier investigators were hampered by false energy, an experimental difficulty; therefore, the transmission of these materials was reinvestigated.[2]

Only recently, Hadni has shown that several of these crystals become quite transparent at liquid-helium temperature.[3] The transmission dependency upon the temperature should be considered when selecting window and filter materials for use at liquid-helium temperatures.

2. For a complete bibliography of transmission of crystals, see E. D. Palik, "A Far Infrared Bibliography," Natl. Res. Lab. Bibliog. No. 21, U. S. Dept. Commerce, Office of Tech. Serv.

3. A. Hadni, J. Claudel, X. Gerbaux, G. Morlot, and J. Munier, Appl. Opt. 4, 487–494 (1965).

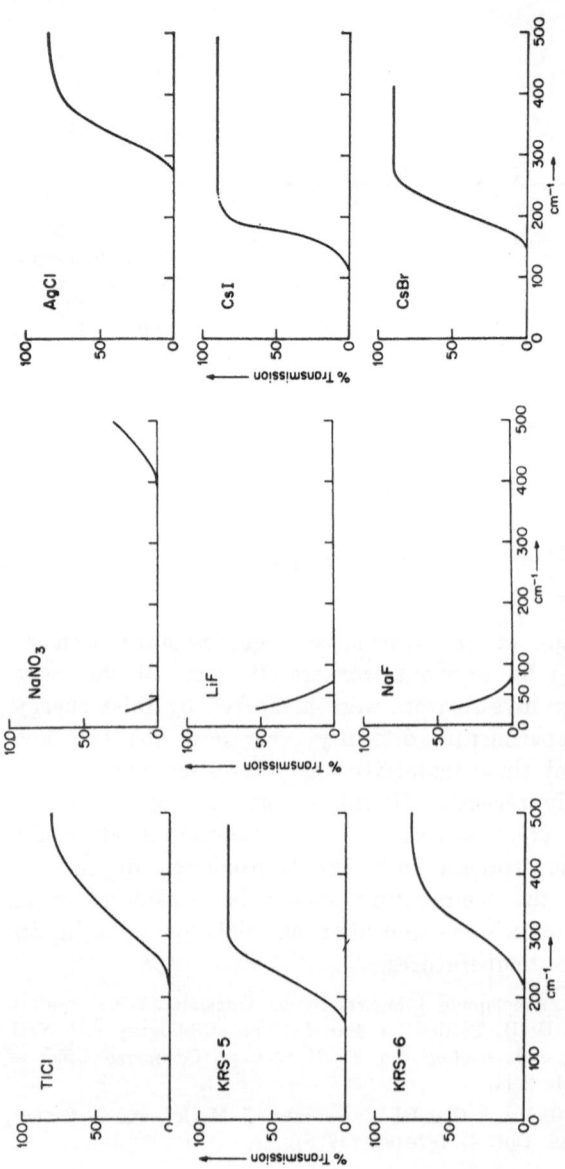

Fig. 6. Transmission spectra of AgCl, CsI, and CsBr, 1 mm thick.

Fig. 5. Transmission spectra of NaNO₃, LiF, and NaF showing low-frequency transmission, 1 mm thick.

Fig. 4. Transmission spectra of TlCl, KRS-5, and KRS-6, 1 mm thick.

The published spectra should help in the selection of filters to eliminate either unwanted shorter-wavelength radiation or, as the case may be, to remove longer-wavelength frequencies. We observed sodium nitrate, lithium fluoride, and sodium fluoride crystals to be opaque in the far-infrared, 400–100 cm^{-1}, but become transparent below 50 cm^{-1}. We have found these to be excellent transmission filters for studies in the region below 50 cm^{-1}.

The transmission characteristic of silicon is much better than crystal quartz. Where quartz has absorption at 128 cm^{-1} and the upper useful transmission limit of approximately 250 cm^{-1}, silicon showed none of these disadvantages. Also, since silicon can easily be cut wedge-shaped, there was no fringing pattern due to internal reflections from parallel surfaces. We felt that silicon windows can be as useful and possibly more practical than high-density polyethylene or quartz because (a) there are no interfering absorptions inherently present in the material, (b) it is resistant to absorption of solvents and chemicals in general, (c) the commercial availability of wedge-shaped windows which do not give interference fringes, and (d) the low cost of the material as compared to crystal quartz.

2.16   Low Temperature Far-Infrared Transmission Properties of Polycrystalline and Single-Crystal Silicon

J. R. Durig, S. F. Bush, and F. G. Baglin

Department of Chemistry, University of South Carolina, Columbia, South Carolina 29208

Fateley *et al.* have reported that the far-infrared spectrum of polycrystalline silicon at room temperature has a 50% transmission curve from 30–500 cm^{-1} [1] We wish to report the appearance of three bands in

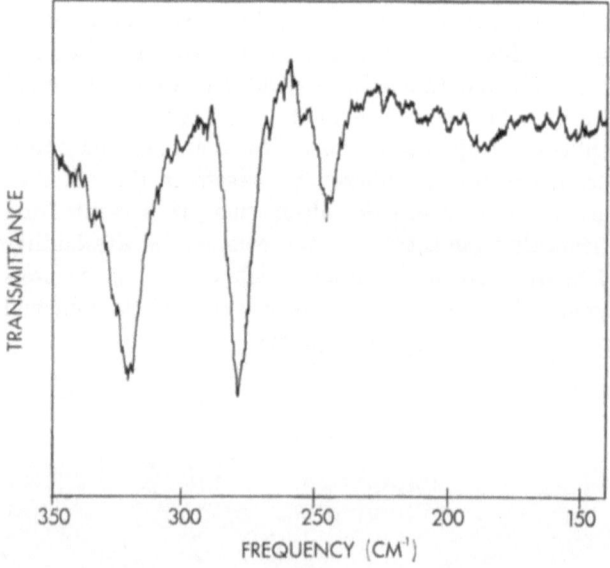

FIG. 1(a). Spectrum of polycrystalline silicon at −196°C.

1. W. G. Fateley, R. E. Witkowski, and G. L. Carson, Appl. Spectry. **20**, 190 (1966).

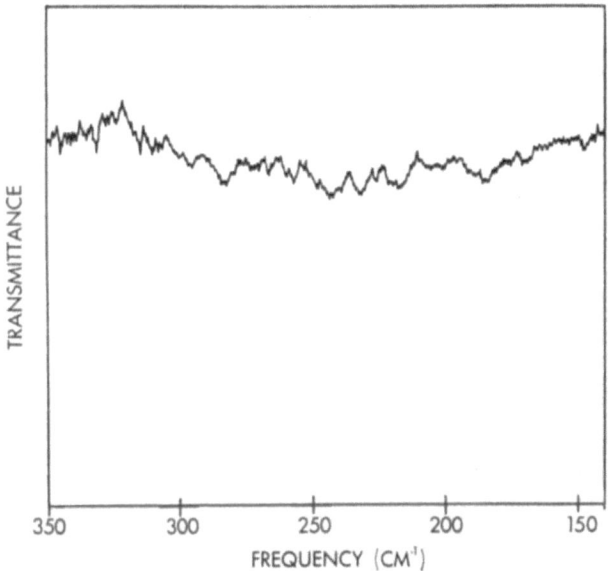

FIG. 1(b). Spectrum of polycrystalline silicon at 25°C.

similar material upon cooling to liquid-nitrogen temperature. A standard low-temperature cell was used for the measurements. A Beckman model IR-11 spectrophotometer and a Perkin–Elmer 521 spectrophotometer were used to record our data.

Upon cooling, bands appear at 245 cm⁻¹, 282 cm⁻¹, and 320 cm⁻¹. The spectra of the polycrystalline silicon is shown in Fig. 1(a) at −196°C, and in Fig. 1(b) at room temperature, for comparison. The bands can be made to appear and disappear as often as the plate is cooled with liquid nitrogen and then warmed to room temperature. We believe that, on the basis of the supplier's information,[2] the impurities in the silicon, i.e., Al, P, B (the order indicates the approxi-

2. Exotic Materials Inc., 2968 Randolph Avenue, Costa Mesa, California.

mate percent of each impurity), have given rise to conduction bands in the silicon spectrum. These bands are not visible at room temperature because the Boltzman factor allows population of all the impurity-electronic levels. However, at low tempeatures the electrons fall into their lowest energy state and discrete transitions can be observed. This phenomenon is well known to solid-state physicists and is identified as the donor-phonon spectrum of, in this case, Al, P, and B dopants in silicon.

Since impurities tend to gather at crystal boundaries, it was hoped that the use of single-crystal silicon would remedy this problem. Unfortunately, it is difficult to obtain single-crystal material in large plates ($2\frac{1}{4}$ by $1\frac{3}{8}$ by 0.05 in. wedged at $1' \pm 30''$ angle). These large plates are needed so that the far-infrared spectrophotometer's beam is not attenuated. For this reason, a small 1-in. disk of single-crystal material was obtained from the supplier.[2] A mid-infrared cold cell and a Perkin–Elmer model 521 spectrophotometer were then used to record the spectrum of the single crystal. No absorption could be detected in the frequency range from 250 to 400 cm^{-1} for the silicon single crystal at either ambient or liquid-nitrogen temperatures.

The polycrystalline material [spectrum shown in Fig. 1(a)], had a very low resistivity (about 0.05 Ω-cm), so material with higher resistivity was sought. To obtain optical-quality silicon for low-temperature studies, one must use type-N silicon with no zone leveling (no dopant), low dislocation (etch pit) density, a resistivity between 70 and 100 Ω-cm, and the 111 plane parallel to the edge. Furthermore, the silicon must be prepared by the Czochralski process to allow sufficiently large crystals to be grown. High quality material of sufficient size, is available for far-infrared studies at low temperatures.[3] A plate with

3. Elmat Corporation, 405 National Avenue, Mountain View, California.

the above specification was cooled to liquid-nitrogen temperatures and the spectrum was recorded; it was identical to that found at room temperature, i.e., no bands were observed between 33 and 700 cm^{-1}.

Acknowledgments—This problem was discovered while one of us (FBG) was working under a National Cancer Institute fellowship, number 6-F2-CA-35,825-01A1. We wish to thank the Exotic Materials Company for the loan of the single-crystal silicon disk. Also, we wish to thank Professor Leconte Cathy for his helpful discussions with us on this problem and the National Aeronautics and Space Administration for financial support by grant NGR-41-002-003-1.

Internal Reflection Spectroscopy in the Far-Infrared Region* 2.17

G. L. Carlson and W. G. Fateley

Carnegie–Mellon University, Mellon Institute,
Pittsburgh, Pennsylvania 15213

This note describes the extension of internal reflection spectroscopy into the far-infrared region below 150 cm^{-1}. To the best of our knowledge this is the first reported use of internal reflection techniques below 250 cm^{-1}, the cutoff point of the widely used KRS-5 reflector plate material.

Spectra were obtained on a Beckman IR-11 far-infrared spectrophotometer utilizing a Wilks' model 9 internal reflection attachment. The internal reflection element was a trapezoidal silicon plate with dimensions 50-×20-×8.5-mm thick and $\theta = 45°$ which was

* The financial support of the Air Force Materials Laboratory under Contract No. F 33 615-68-C-1369 is gratefully acknowledged.

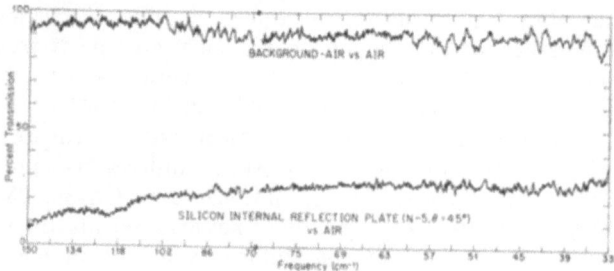

FIG. 1. Far-infrared transmission of silicon internal reflection element with N = 5, θ = 45° relative to air.

specially fabricated by the Wilks Scientific Corp. from a blank of single crystal silicon[1] kindly provided by the Dow Corning Corporation.

The 8.5-mm crystal thickness was chosen to accommodate the large beam size and wide slits normally required in the far-infrared region. The number of reflections within this silicon trapezoidal reflector plate is approximately 5 as calculated from the equation $N = $ (length/thickness) $\cot\theta$.[2] This indicates that the radiation must pass through about 60 mm of silicon, and yet the reflector plate shows reasonable transmission below 100 cm^{-1} after taking into account the reduction in energy by the internal reflection attachment and the high reflection loss at both air–silicon interfaces. The transfer optics of the Wilks' model 9 accessory transmit about 60% in the far-infrared region so the transmission of the silicon crystal in the model 9 attachment should be approximately 30% if the only attenuation by the crystal was due to reflection losses. As shown in Fig. 1, the transmission approaches this 30% value below 100 cm^{-1}, but falls

1. The single crystal silicon was characterized as float zone, 111 orientation, phosphorus-doped N-type with resistivity >300 Ω cm.

2. N. J. Harrick, *Internal Reflection Spectroscopy* (Wiley–Interscience, Inc., New York, 1967), Chap.2.

off to a transmission of only 5% at 150 cm⁻¹. This drop in transmission must be due to absorption by the crystal which restricts its range of usefulness to the region below 150 cm⁻¹. It may be possible through the use of very high resistivity silicon to extend this range to 250 cm⁻¹ and overlap the lower limit of the KRS-5 internal reflection element.

An estimation of the depth of penetration and effective path length achieved with the 5-reflection, silicon internal reflection plate was made using the equations given by Harrick.[2] For a crystal index of 3.4[3] and a sample index of 1.5, the depth of penetration at 200 μ (50 cm⁻¹) for an angle of incidence of 45° is calculated to be approximately 0.02 mm which would correspond to an effective path length of 0.10 mm for a 5-reflection

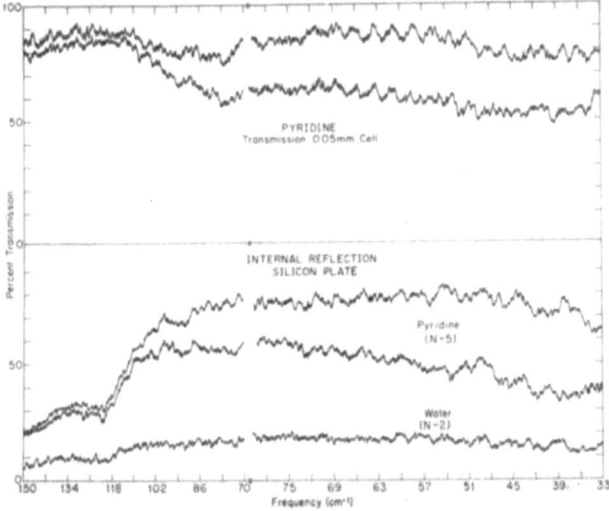

FIG. 2. Top: Far-infrared transmission spectrum of pyridine in a 0.05-mm polypropylene cell. Bottom: Far-infrared internal reflection spectra of pyridine and water utilizing a silicon internal reflection element with $\theta = 45°$ and N as indicated.

3. C. M. Randall and R. D. Fawcliffe, Appl. Opt. 6, 1889 (1967).

element. This value can be verified by comparing the low frequency internal reflection and transmission spectra of liquid pyridine shown in Fig. 2. Unfortunately, we do not know of any liquid samples which have discrete low frequency absorption bands that are detectable at a path length of 0.10 mm; thus it was necessary to make this comparison using the strong general absorption of pyridine. Figure 2 also gives the internal reflection spectrum of liquid water. Water has a very intense spectrum in the far-infrared region, and it was necessary to place the sample on only one side of the silicon internal reflection element. Therefore this spectrum corresponds to an effective path length of approximately 0.04 mm.

Figure 3 gives the far infrared internal reflection and transmission (Nujol mull) spectra of solid HgS, and the internal reflection spectra of CsI and benzimida-

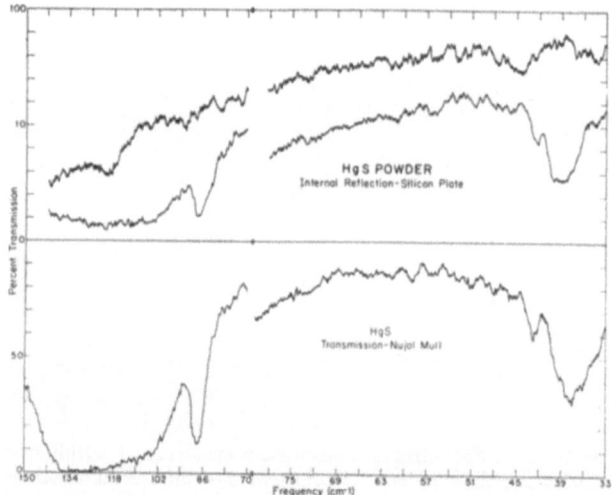

FIG. 3. Far-infrared internal reflection ($N = 5$, $\theta = 45°$) and transmission spectra of solid HgS.

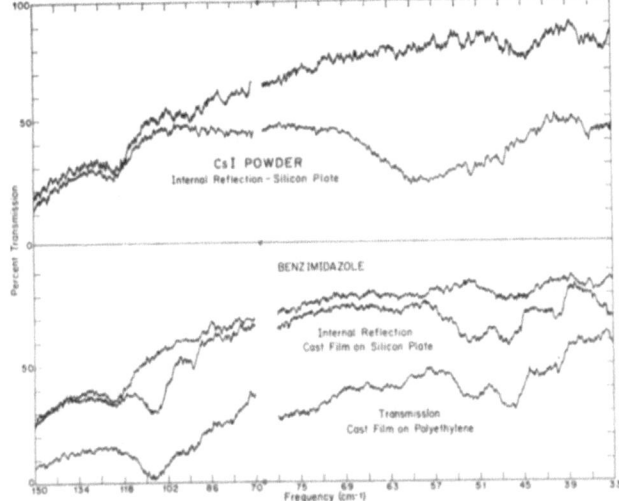

Fig. 4. Top: Far-infrared internal reflection spectrum of CsI powder. Bottom: Far-infrared internal reflection and transmission spectra of cast films of benzimidazole. For the internal reflection spectra, $N = 5$, $\theta = 45°$.

zole are shown in Fig. 4. In the case of HgS and CsI, the powered samples were simply dusted onto both sides of the silicon plate while for benzimidazole a film of the solid was deposited on the plate from ethanol solution.

It would appear that, at this stage of development, internal reflection techniques in the far-infrared region will not have wide application. Most samples require path lengths of at least 1.0 mm for the region below 150 cm^{-1} and an effective path length of 0.1 mm will drastically limit the number of compounds that will be amenable to study by this technique. However for strongly absorbing solids and liquids, the ease of sample preparation may make the method quite desirable. It is also possible that by using the crystal in a double-sampling mode or reducing the thickness of the silicon crystal to some optimum size which would give a larger number of reflections while not greatly wasting instrument energy could as much as double the effective path length. It may also be possible with higher purity silicon and with an identical, compensating unit for the reference beam of the spectrophotometer to extend the useful range to 250 cm^{-1} and thus overlap the region accessible with the KRS-5 internal reflection element.

ACKNOWLEDGMENT

We wish to thank John Baker of the Solid State Division, Dow Corning Corp. for the gift of the silicon blank, and Paul Wilks of the Wilks Scientific Corp. for the special fabrication of the internal reflection element.

2.18   Solution for the Sample Contact Problem in ATR

Tomas Hirschfeld*

Faculty of Chemistry, Montevideo, Uruguay

To obtain good attenuated total reflection (ATR) spectra,[1-3] very good contact between the samples is required. Solid samples often have surfaces irregu-

---

*Presently on leave of absence at North American Aviation Science Center, 1049 Camino Dos Rios, Thousand Oaks, Calif.

1. J. Fahrenfort, Spectrochim. Acta **17**, 698–709 (1961).
2. J. Fahrenfort. Proc. 10th Colloq. Intern. Spect., Maryland, 437–60 (1962).
3. J. Fahrenfort and W. M. Visser, Spectrochim. Acta **18**, 1103–1116 (1962).

lar enough to require them to be pressed against the prism, to improve contact by deformation.[4]

If the sample is harder or tougher than the prism material,[5,6] this procedure leads to rapid deterioration of the prism. Unfortunately, the high index materials used for ir ATR are not very strong; and, while much harder ones are available for uv ATR, the lower wavelengths require a much better contact.

One may increase penetration and lower contact requirements by working at incidence angles approaching the critical one. However, this leads to severely ''distorted'' spectra that do not follow Beer's law, and consequently, are of limited value.[7]

Currently, micro ATR equipment is used for these samples, in which the beam is focused on the specimen's surface.[8] Here one can select the best parts of the surface for analysis, but this is useful only if the surface's properties are uniform. Also, the wide spread of incidence angles generally results in failure of the obtained spectra to follow Beer's law.

Multiple-reflection ATR improves sensitivity enough to reduce contact quality requirements,[9] but large samples are required and, even then, it is difficult to observe the weaker bands.

4. R. L. Harris, ''Factors Affecting the Reproducibility of ATR Spectra,'' Pittsburgh Conf. Anal. Chem. and Appl. Spectry. (1964).
5. J. F. Deley, R. J. Gigi, and A. J. Liotti, J. Tappi 46, 188–92A (1963).
6. C. F. Yelin, ''Identification of Wire Enamels by IR Internal Reflection Spectroscopy,'' Anachem. Conf., Detroit. (1965).
7. S. E. Polchlopek, Appl. Spectry. 17, 1124 (1963).
8. Connecticut Instrument Co., CIC Newsletter No. 20.
9. P. J. Wilks and M. R. Iszard, ''The Identification of Fibers and Fabrics by Internal Reflection Spectroscopy,'' Develop. Appl. Spectry. 4, 141–149 (1964).

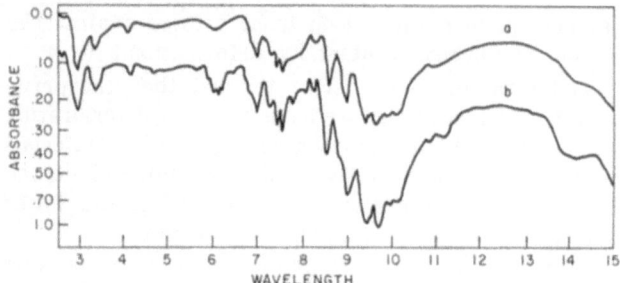

Fɪɢ. 1. ATR spectra of cotton fabric with and without deformable solid contacting.

Various authors[3,10-12] have used a contact fluid between the sample and the prism. With soft or thin samples, this may actually bring the surfaces closer together.[5] However, for hard samples with surface irregularities, a contact fluid will bring improvement only when its index is so high as to preclude total reflection on the prism–fluid interface. Liquids with a wide transmission range and a high enough index (more than 2 for Ge at 45° or KRS-5 at 60°) are not available. The use of liquids of lower index does not improve sensitivity,[11] and in fact, rather degrades it.[12]

In the cases in question, the essential characteristic of the intermediate layer is not fluidity but easy deformability. Therefore, for this application it is possible to use the available soft high-index ir transmitting materials (AgCl or AgBr, for example). The sample is then pressed into a small thin sheet of this material held against a smooth nonadherent backing, and the whole transferred to the ATR accessory with the smooth face against the prism. Very slight pressure will then result in near-perfect contact.

10. B. N. Ingram, Res. Develop. No. 34 **18** (June 1964).
11. R. Bent and W. R. Ladner, Fuel **44**, 243–724 (1965).
12. T. Hirschfeld, Can. Spectry. (to be published).

Because of scattering and possible violation of the total reflection condition, the method will work only if the irregularities present are not very steep. Curve a in Fig. 1 shows the ATR spectrum of a cotton fabric on a KRS-5 internal-reflection element giving five reflections on the sample at a 45° angle. The spectrum was recorded on a Perkin–Elmer 337 spectrophotometer.

Curve b shows the same sample after pressing into a 0.5-mm rolled AgCl plate (Harshaw Chemical Co.) that was then contacted to the prism with a slight pressure. The intensity of the spectrum can be seen to have more than doubled by this procedure, the increase being more marked towards shorter wavelengths.

Other contacting media of high enough index would be the semiliquid (at ambient temperature) As–S–I ir-transmitting glasses recently described by Flaschen[13] or, in the uv–vis region, some of the ruby laser immersion liquid solutions described by Graham.[14]

Method for Obtaining Improved Diffuse Reflectance Spectra in the Near Infrared

2.19

A. F. Sklensky, J. H. Anderson, Jr., and
K. A. Wickersheim

Materials Science and Mineral Engineering Departments,
Stanford University, Stanford, California 94305*

In the course of an investigation of adsorbed water and hydroxyl groups, we wished to obtain near-in-

13. S. S. Flaschen, A. P. Pearson, and W. R. Northover, J. Appl. Phys. 31, 219 (1960).
14. N. E. Graham, B. I. Davis, and D. V. Keller, Appl. Opt. 4, 613–615 (1965).

*The authors' present address is Lockheed Research Laboratory, Palo Alto, California 94304.

frared spectra from silica powders. Although it was
clear that diffuse reflectance techniques would satisfy
our requirements, no combination of spectrometer
and reflectance accessory suitable for use in the
range of principal interest $(1-3\,\mu)$ was available to
us. Because we had access to a Cary Model 14 RI
spectrophotometer, it seemed most reasonable to consider modification of the reflectance accessories available for this instrument. The purpose of this note is
to describe the approach adopted, the results achieved,
and the possibilities of improving the system further.

A number of diffuse reflectance accessories are
available from the manufacturer for the Model 14
spectrophotometer. These accessories are intended for
use in the visible and ultraviolet portions of the
spectrum. Most units substitute for the entire sample
compartment casting. However, one accessory consists of small integrating spheres which fit within the
normal sample cell spaces. (See Fig. 1.) While the
"in-cell-space" accessory is optically the least efficient of the available reflectance accessories, we
chose to adapt it for our purposes because of its lower
cost and because its use allows easy changeover from
reflection to transmission modes of operation.

Preliminary information[1] indicated that the "in-
cell-space" spheres could not be used for our purposes
without modification because the optical surfaces are
coated with a barium sulfate paint. This coating ad-
sorbs water and therefore absorbs strongly in the
vicinity of 1.4, 1.9, and 2.7 μ, reducing the transmission of the spheres below the usable level in the
regions of greatest interest to us.

Our first attempt at improving the infrared performance of the spheres was to remove the barium
sulfate coating and use the bare metal (aluminum)
surfaces. The resulting signal was barely usable, how-

---

1. J. R. Eno, Applied Physics Corp. (private communication).

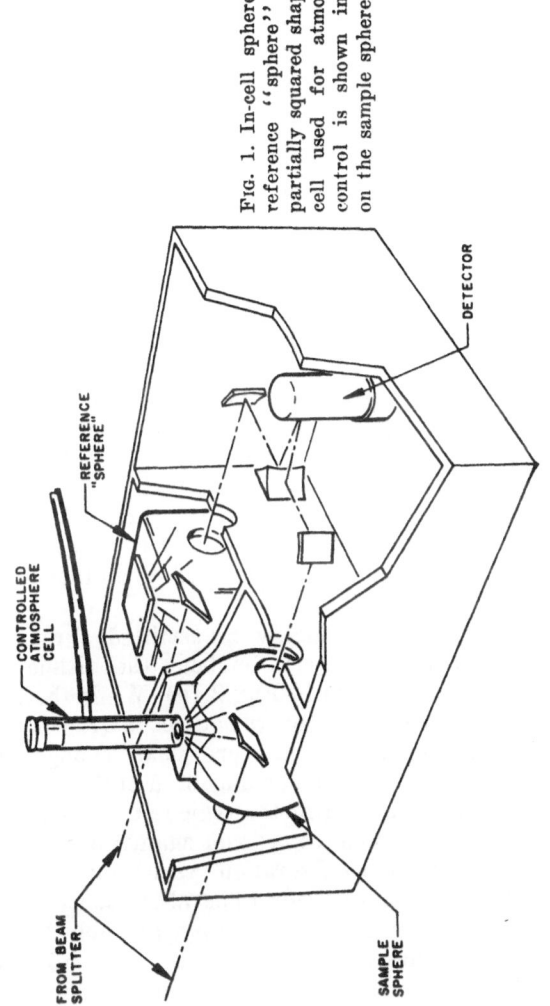

FIG. 1. In-cell spheres: the reference "sphere" has a partially squared shape. The cell used for atmospheric control is shown in place on the sample sphere.

CONTROLLED ATMOSPHERE CELL

REFERENCE "SPHERE"

FROM BEAM SPLITTER

SAMPLE SPHERE

DETECTOR

ever, and a more efficient coating was sought. Finely ground alumina was found to give good results in both the near infrared and visible, and this is the coating we adopted.

The procedure developed for applying this coating is as follows: Before applying the alumina, all optical surfaces inside the spheres are cleaned thoroughly, first with scouring powder and then with Alumiprep. All nonoptical surfaces (and mirrors) are masked. A water slurry thin enough to spray, and containing about 0.01% CMC as a binder, is mixed in a beaker. The slurry is applied to the spheres with a siphon-type spray gun using a flexible tube to convey the slurry from the beaker to the spray gun. Settling of the slurry in the beaker is prevented by magnetic stirring.

For the best results, very thin coats must be applied, each being allowed to dry completely before the next is applied. Five to eight coats are adequate. Reagent-grade alumina adheres best, probably because of the random sizes of the particles. After the final coating has dried, very fine (0.3 to 1μ) polishing-grade alumina may be added (dry) to the surface to give a more uniform finish.

The spectrum of energy transmitted through a single, recoated sphere throughout the visible and near infrared is shown in Fig. 2 (a). While there are still some absorptions (or variations in transmission of energy) observed, these are small enough that energy adequate for the taking of useful spectra is available through most of the region of interest.

When the two spheres are run against one another in a double-beam configuration, variations in transmission are still observed. (The absorptions appear as peaks in the transmitted energy, probably because of a greater number of reflections in the distorted reference sphere.) The variation in reflection with wavelength is small and may be eliminated by the

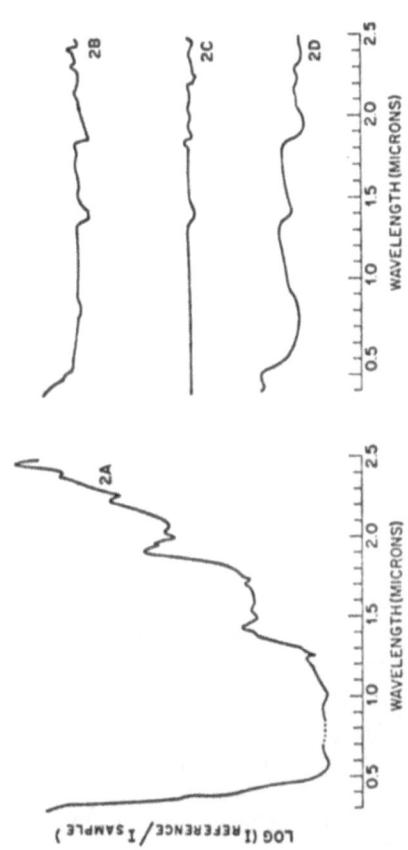

FIG. 2. Spectra of sphere coating. (a) Absorption spectrum of a single sphere. (b) The sample sphere run versus the reference sphere, with an alumina slurry on a highly reflective backing covering the sample port, and loose alumina, very thick, covering a window on the reference port. (c) The same configuration as (b) with the baseline adjusted for maximum flatness. (d) With the baseline adjusted as in (c), the sample port is here covered by an alumina slurry on an absorbing backing. Comparison of band d reflects the opacity of the coating. The approximate useful range of the instrument can be seen to be 0.3–2.5μ.

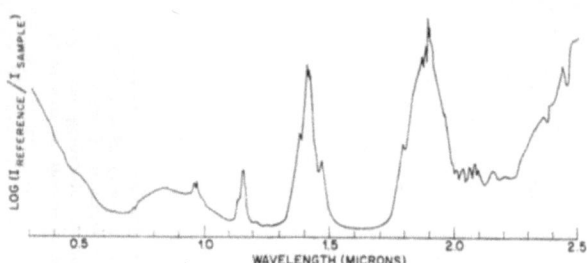

FIG. 3. Beryl: a rough-cut sample placed on the sample port. Spectrum was found to be independent of crystallographic orientation.

normal baseline compensation procedure described in the instrument manual. The double-beam spectrum before and after compensation is shown in Fig. 2, Curves (b) and (c), respectively.

In order to check the opacity of the coatings so obtained, two substrates, one highly reflecting and one highly absorbing, were coated with alumina by our technique. These were then run as samples using the finished spheres. The results are shown in Fig. 2 (d).

Performance of the system can perhaps be best judged by a consideration of reflectance spectra actually obtained. Spectra have been obtained using three different sample configurations.

1. The slab or pressed-pellet technique: An optically thick slab of material or a pressed pellet may be placed over the sample port of the sphere. The beryl spectra, shown in Fig. 3, were obtained from a rough-sawn section of a large, translucent crystal. The orientation (polarization) effects observed in transmission[2] are not observed in the reflectance spectra.

2. Window or external cell techniques: A powdered sample can be spread on a transparent window

2. K. A. Wickersheim and R. A. Buchanan, Am. Min. **44**, 440 (1959).

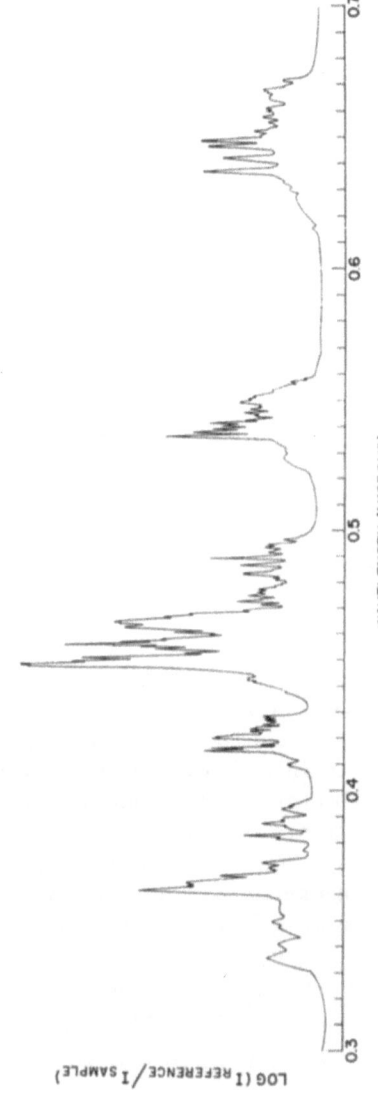

FIG. 4. Holmium oxide: demonstrating resolution in the visible; the sun gun attachment is the light source, and a 1P28 photomultiplier is the detector.

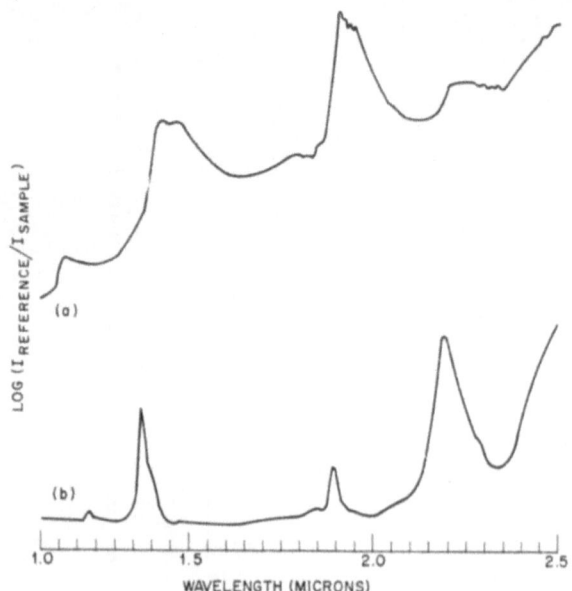

FIG. 5. Silica gel: (a) as received, and (b) after being subjected to vacuum for a week.

or in the bottom of a flat-bottomed cell, these in turn being placed on the sample port. The spectrum of holmium oxide, shown in Fig. 4, was obtained from powder spread on an Infrasil window. If the sample cell is closed, atmosphere control can be achieved. For example, the spectra of powdered silica gel obtained in such a cell, first at atmospheric pressure and then under vacuum, are shown in Fig. 5. (A discussion of the silica gel spectra is given by Anderson.[3])

3. Test tubes or internal cell techniques: When the sample exhibits small absorptions, making mul-

3. J. H. Anderson, Jr. and K. A. Wickersheim, Surface Sci. 2, 252 (1964).

tiple reflections from the sample desirable, the sample may be poured into a tube or cell and lowered into the interior of the sphere via the sample port.

The spectra presented are of weakly absorbing materials and resemble absorption spectra, that is, we detect the loss of energy due to absorption within the sample rather than the coincident increase in the reflection from the sample. Particle sizes were of the order of microns (in the rare-earth oxides and silica gel) and millimeters to centimeters in beryl. The dependence of reflectance spectra on particle size is treated in detail by Kortum.[4]

A variety of spectra have been presented to demonstrate the performance and versatility of the system. Much of the versatility derives from the sample-handling advantages of the reflectance technique itself. The performance of the Model 14 RI in the near infrared is, of course, considerably better than that of a conventional Model 14 spectrophotometer, and performance has been further improved through the use of the high-intensity source (which uses a quartz iodine ''Sun Gun'' lamp) sold as an accessory to the Model 14. Nonetheless, some useful extension of reflectance techniques into the infrared should be possible with our spheres and a conventional Model 14.

An obvious improvement would be to apply the alumina coatings to one of the optically more efficient reflectance accessories manufactured by Applied Physics Corporation. One would expect an order of magnitude increase in optical efficiency[1] allowing an increase in both resolution and range.

Another possible modification which could be of use when the absorptions of interest are very weak, as might be the case in the study of surface films on nonporous materials, would be to coat the entire in-

4. G. Kortum, W. Braun, and G. Herzog, Angew. Chem. Intern Ed. Engl. **2**, 333 (1963).

terior of the sample sphere with the sample material using the technique described herein. The spheres could also be equipped with windows to allow atmosphere control over the entire reflecting surface. We have not attempted such experiments as yet.

We wish to thank Professor G. Parks for his help and stimulation and J. R. Eno and R. C. Hawes of Applied Physics Corporation for their advice and assistance. The work was supported by the Stanford Center for Materials Research and by the Lockheed Independent Research Fund.

2.20   Calibration of Infrared Cells

A. M. Burrill

The Procter and Gamble Company, Ivorydale Technical Center, Cincinnati, Ohio 45217

To minimize the error caused by cell thickness changes, we have made it a practice to recalibrate our cells frequently by scanning ten to twenty interference fringes in triplicate and calculating the thickness from the equation

$$t = n/2(f_1 - f_2). \qquad (1)$$

This was a costly operation, and it occurred to us that, since the interference pattern is a function of cell thickness, the frequencies at which the transmission maxima occur are also functions of the cell thickness. Thus we should be able to measure the cell thickness from the transmission maximum of a single interference fringe.

It can be shown[1] that, whenever the cell thickness t is an even increment of a quarter wavelength, a transmission maximum occurs in the interference pattern. Expressed mathematically, the transmittance T is a maximum whenever k in the expression

$$t = 2k_\lambda/4 \qquad (2)$$

is a whole number.

If we could scan the interference pattern of a cell from two microns to infinity, where the cell thickness would be negligible compared to the wavelength of light, we would have a transmission maximum at infinity $(k=0)$. The next would appear when the wavelength of light was twice the cell thickness $(k=1)$. A third maximum would occur when the wavelength of light was exactly equal to the cell thickness $(k=2)$. Each transmission maximum in the interference pattern has its own integral value of k. If we determine cell thickness by some procedure we can calculate the k value for any maximum. Small errors in the original calculation of cell thickness will cause k to be other than an integer, but rounding off to the nearest whole number will result in the correct k value. Using the correct k value we can go back and recalculate the cell thickness. In practice, the accuracy obtained with this technique is several times better than that obtained by the conventional technique.

The procedure we now use in our laboratory is as follows: Place the clean, dry cell in the sample holder of the spectrometer. Scan the spectrum at 100 to 200 cm⁻¹ per minute. With a properly prepared cell, the

1. F. W. Sears, *Principles of Physics III, Optics* (Addison-Wesley Publ. Co., Inc., Cambridge, Mass., 1946), pp. 150–157.

interference pattern is a series of sine waves on the recorder chart.

For instruments linear in frequency, the cell thickness is approximated from the equation

$$t=n/2(f_1-f_2) \qquad (3)$$

where t=cell thickness in centimeters, f_1 and f_2=frequency in cm^{-1} bounding the portion of the interference pattern chosen for measurement, and n = the number of full waves between f_1 and f_2 on the interference pattern.

To determine cell thickness accurately, select a point of maximum transmission on the interference pattern. Draw a horizontal line approximately halfway between the maximum and minimum transmission. (See Fig. 1.). Carefully measure the frequency at which the horizontal line intersects the sides of the peak. This gives a more accurate estimate of f_0 than can be obtained by measuring f_0 directly.

For instruments linear in frequency determine the k value of the peak selected, using the equation

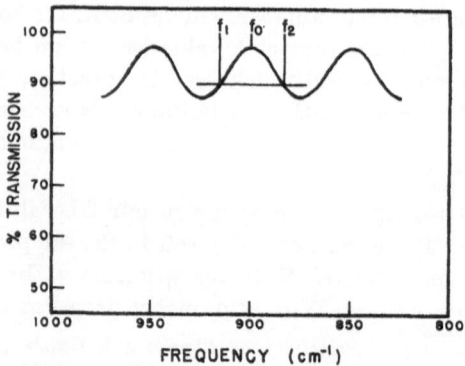

Fig. 1. % transmission vs frequency (cm^{-1}).

$$k=2tf_0 \tag{4}$$

$$k=t(f_1+f_2), \tag{5}$$

where t=cell thickness in centimeters as previously determined, f_0=frequency of maximum transmission, f_1 and f_2=frequency in cm^{-1} at which the horizontal line intersects the sides of the peak, and k=a whole number.

If k is not a whole number, round it off to the nearest whole number. Having determined the k value of a peak, a more accurate cell thickness may be determined by solving the equation

$$t=k/(f_1+f_2) \tag{6}$$

for t.

Any error in the frequency calibration of the instrument will result in an error in the measurement of cell thickness. Correct this error by scanning near the interference fringe a polystyrene peak for which the k value is being determined. The polystyrene is scanned on the same chart paper without removing the chart paper from the drum. Determine the difference, Δf, between the measured frequency and the true frequency from

$$\Delta f=f_{\text{true}}-f_{\text{observed}}. \tag{7}$$

Correct f_1 and f_2 by adding Δf to the measured frequency.

After determining cell thickness, check it at frequent intervals to prevent error from occurring due to changes in cell thickness. If the k value and frequency of the transmission peak are recorded, these will identify the peak when rechecking the cell thickness.

There are two limitations to this technique. The first is that the larger the k value of the fringe, the more accurate must be our initial estimate of cell thickness; otherwise, an error will occur in the determined k value. The second limitation is that a linear displacement along the frequency scale will cause a greater error by the new technique than by the old. This should not be a serious problem, however, with modern grating instruments.

2.21   Polishing Jig for Infrared-Cell Windows*

W. R. Feairheller, Jr. and H. DuFour

Monsanto Research Corporation, Dayton, Ohio

Frequent hand-polishing of infrared-cell windows, particularly the softer materials such as cesium iodide and bromide, can easily result in a convex surface and a wedge shape instead of flat, parallel sides. A simple jig has been used in this laboratory for some time to prevent this occurrence during grinding of the crystal blanks and polishing to final finish.

The device, shown in Fig. 1, consists of simple turnings easily made on any lathe. Our jig was made with leadalloy, a steel containing a small amount of lead for machinability. As we normally use two sizes of cell windows, 25 and 32 mm diam, the ends of the jig were designed to accommodate both window sizes. A convenient size for the barrel was found to be about 2 in. in diameter by $1\frac{3}{4}$ in. high.

---

*This work was supported by the Materials Laboratory, Research and Technology Division, U. S. Air Force, under contract.

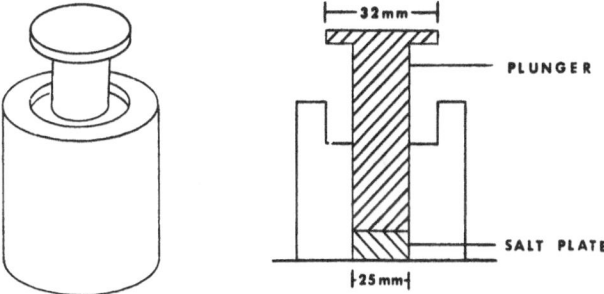

Fig. 1. Perspective and cross-sectional views of infrared-cell-window polishing jig.

For grinding the cell blanks, additional pressure is exerted on the plunger until the major scratches are removed. For final grinding and polishing operations, the plunger has sufficient weight (for the softer materials) and no additional pressure is required. It is important to remove any traces of the grinding compound from the cell window or jig before attempting fine grinding and polishing, or new scratches will be produced on the surface of the salt plates.

The jig can easily be altered to fit any size of round cell windows, and the barrel can be milled to hold square or rectangular plates, as required.

Additional advantages of the jig are that finger prints and finger moisture are not a problem, and use of the jig prevents build-up of the polishing compound on the edges of the windows.

2.22   Restoration of Transmission to Silver Chloride Plates

Donald Trimnell

Northern Regional Research Laboratory,
U. S. Department of Agriculture,
Peoria, Illinois 61604

Silver chloride plates used in infrared spectroscopy often become tarnished and darkened when exposed to light, air, and sulfur compounds. Moreover, with continued use scratches are formed, which reduce the transmission, particularly at the shorter wavelengths.

Present methods of restoring transmission to such plates involve either a chemical action upon the surface to dissolve the tarnish or a mechanical-grinding operation to restore the surface.

The following method combines the advantages of chemical and mechanical action, is simple and easy to use, and involves a minimum of toxicity hazards. A 1:1 mixture of commercial silver and copper polishes (The International Silver Company, Meriden, Connecticut[1]) is dabbed onto a cotton swab and rubbed vigorously over both sides of the plate. The excess polish is removed either by buffing with a clean cotton swab or by rinsing with distilled water and buffing. The process may be repeated several times to restore maximum transmission. The cleaning process may be automated if desired by use of motor-driven buffers impregnated with the polish. This method both removes tarnish and eliminates scratches, thus restoring transmission and increasing the effective life of the plates.

1. Mention of trade names or commercial products does not imply endorsement by the U. S. Department of Agriculture over similar trade products not mentioned.

High Photometric Accuracy in an Optical Null, Double-Beam Infrared Spectrophotometer*

2.23

K. E. Stine and R. A. Weagant†

Beckman Instruments, Inc., Fullerton, California 92634

To date, optical null infrared spectrophotometers have been capable of double-beam photometric accuracies of no better than 0.5% to 1.0%. When operated in the double-beam mode, optical null instruments measure the transmittance of a sample by optically attenuating the reference beam until its energy equals that of the sample beam. This operation is usually performed automatically by a servosystem which constantly repositions an optical attenuator (comb) in the reference beam to obtain a null balance condition. A pen servosystem then follows the comb movement from a linear potentiometer geared to the comb.

When the transmission of a sample is determined by comparison to an optical attenuator, it is generally assumed that the transmitted energy is directly proportional to the attenuator position. Unfortunately, this is not always true. Because of imperfections in the comb shape and beam inhomogeneties in the comb area caused by the nonuniform surface of the source, the position of the comb does not always reflect the true sample transmission. This is especially true at very high or very low transmission values where the comb is used at its extreme posi-

---

*Presented in part at the 1968 Pittsburgh Conference on Analytical Chemistry and Applied Spectroscopy.
†Present address: Honeywell Radiation Center, Lexington, Massachusetts.

tions. Relatively large errors frequently occur below the 10% T level.

In single beam operation, where the comb is not used, or by comparing double beam readings with those obtained in single beam, photometric errors can frequently be reduced to 0.2%. Photometric values obtained using the former approach are subject to source and atmospheric fluctuations and, because of energy variations, must usually be taken point-by-point even over short frequency ranges. The second approach, while obviating the problems associated with single-beam measurements, requires that single-beam/double-beam values be compared rigorously to establish a calibration curve.

To optimize an optical null infrared instrument for highly accurate photometric measurements and to minimize the difficulties associated with this type of measurement, a new electromechanical photometric accuracy calibration system has been developed. This system allows double-beam accuracies of better than 0.2% to be obtained dynamically over wide frequency ranges without the aid of a working calibration curve.

System description—The photometric accuracy calibration system linearizes the Beckman IR-4 type of optical null instruments by building into the comb potentiometer an electrical function which corresponds to the true transmission characteristics of the optical attenuator. Basically, the system consists of a special comb potentiometer, a series of trimming potentiometers, and a power supply.

The standard comb potentiometer is replaced by a special 22-tap potentiometer located directly below the comb servomotor, as shown in Fig. 1. Eighteen taps are located at approximately 5% increments within the 10%–95% T range of the potentiometer. The remaining 4 taps are located at smaller increments (2.5%) within the 0–10% T range where the

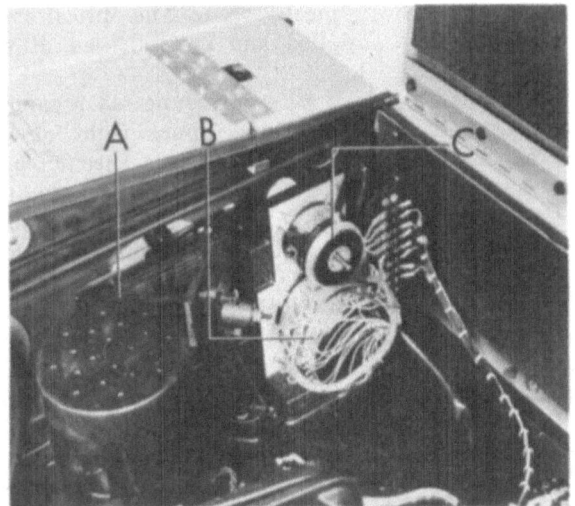

FIG. 1. Comb servosystem: (A) comb, (B) special multi-tapped comb potentiometer, and (C) comb servomotor.

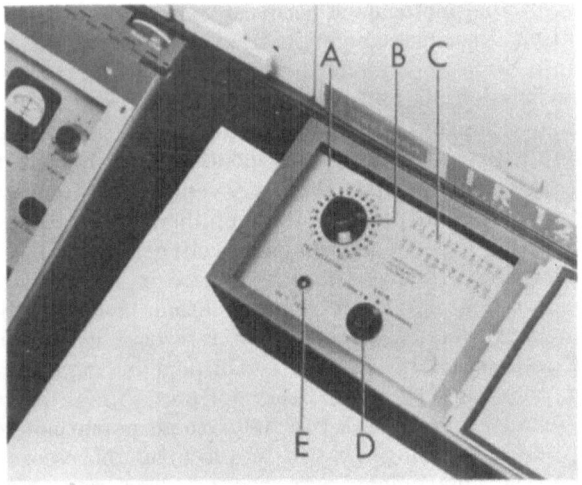

FIG. 2. Photometric accuracy calibrator; (A) control panel, (B) tap selector switch, (C) trimming potentiometers, (D) mode switch, and (E) 100% test button.

largest comb errors typically occur. The optical comb is that normally supplied and can be seen at the left-of-center, Fig. 1.

The wiper of each of 22 series-connected trimming potentiometers, located at the upper right of the control panel, Fig. 2, is connected to a specific comb potentiometer tap. Because the impedance of the series-connected trimming potentiometers is much lower than that of the comb potentiometer (approximately 200 Ω compared to 10 kΩ), the normal voltage at each tap location on the comb potentiometer can be adjusted approximately ±2.5% from normal. This trimming capability allows the comb potentiometer voltage to be matched to the actual energy transmitted by the optical comb over its entire transmission range. A 11-V power supply is added to provide the voltage across the comb and trimming potentiometers, and the pen potentiometer.

Double beam calibration—An understanding of the calibrating procedure sheds some light on the function of the photometric accuracy calibrator. For calibration the instrument is used alternately in the single beam and double beam modes. First, the true transmission of the comb must be determined at each tap location on the comb potentiometer. This is accomplished by moving the **mode** switch, Fig. 2, to its **calibrate** position. This reverses the demodulator phase and presents the amplifier output, rather than the comb potentiometer voltage, to the pen servosystem. When the instrument is switched to single beam operation (sample beam blocked), the pen then records the energy of reference beam. The single beam 100% level is established by depressing the momentary 100% **test** button, which allows the comb servo to seek the 100% comb potentiometer tap, and adjusting the pen to 100% on the recorder with the instrument gain control. The true comb transmission can then be determined at each tap location by moving the **tap selector** (Fig. 2) switch to the tap location of interest and noting the recorder pen position.

Following each single beam reading of comb transmission, the instrument is returned to double beam operation (both beams open) and any discrepancy between the double beam pen position and the true comb transmission is trimmed out with the appropriate trimming potentiometer. This procedure is repeated for each comb potentiometer tap.

The calibration procedure is quite straightforward and, relative to that of earlier devices we constructed, has been simplified by the inclusion of circuitry whereby the comb potentiometer automatically seeks the location of each potentiometer tap. As shown in Fig. 2, the control panel is conveniently located outside the source compartment. Consequently, there is no need to break the instrument purge during calibration or operation of the photometric accuracy calibrator. In practice, calibration of the entire 0–100% T range can be accomplished in 10–15 min. No calibration curve is required since the correction function is electrically built into the comb potentiometer.

Double beam photometric accuracy—To evaluate the system, double-beam accuracies for an IR-12 grating spectrophotometer were determined by comparing double-beam %T values both to single-beam and to linearity disk[1] values. To improve %T readability and eliminate uncertainties due to paper or metal scales, a 1050 count, 0.1% shaft-angle encoder[2] was installed on the shaft of the pen potentiometer. A digital display system[2] was coupled to the encoder for convenient readout of %T values. Using

1. Research and Industrial Instruments Company (Beckman), London, England.
2. Datex Division, Conrac Corporation, Los Angeles, California.

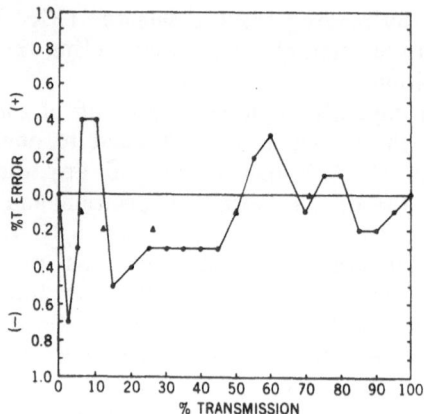

Fig. 3. Uncalibrated double-beam photometric accuracy; by comparison to single-beam (●) and linearity disk (▲) values.

this system, $\%T$ values were found to be readable and repeatable to within 0.1%.

The double-beam photometric error of an IR-12 taken from the production floor, i.e., with no photometric calibration, is shown in Fig. 3. The circled data points (●) and triangular data points (▲) were established by comparison of double-beam readings to single-beam and linearity disk values, respectively. The uncalibrated photometric accuracy of this instrument is well within the published specification of 1.0%. Above the 10% T level, the error is only 0.2%–0.3%. As expected, the worst errors occurred below the 10% T level where the maximum error of 0.7% was found at the 2.5% T level.

Of special interest are the data points (▲) established with linearity disks. If these points had been considered alone, without the aid of single-beam measurements, the photometric error at the worst point would have appeared to be only 0.2%. In this particular case, photometric tests with linearity disks

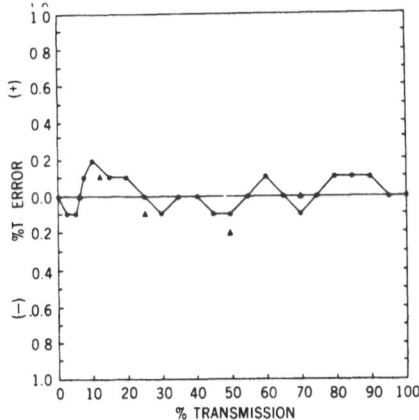

FIG. 4. Calibrated double-beam photometric accuracy; by comparison to single-beam (●) and linearity disk (▲) values.

alone might have led to erroneous conclusions, especially regarding the region below 15% T. It was only coincidental that the linearity disk values fell within those regions of high photometric accuracy, e.g., at 6% T.

The double beam photometric error after calibration of the comb with the photometric accuracy calibration is shown in Fig. 4. The error was reduced to 0.2%, or less, throughout the entire 0–100% T range.

To assess the effect of large slitwidth changes on the photometric accuracy, the error was redetermined under new slitwidth conditions. The preceding tests were performed at 500 cm⁻¹ using a 3.0-mm mechanical slitwidth; appropriate for routine recording of spectra. Without changing other instrument controls, the frequency was changed to 1865 cm⁻¹. The slitwidth was allowed to follow the routine slit program and came to rest at 0.30 mm, thereby

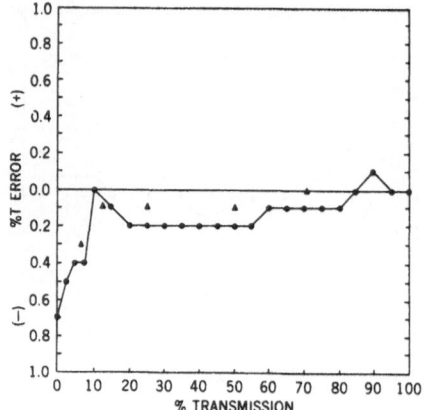

Fig. 5. Calibrated double beam photometric accuracy following 10:1 slitwidth change; by comparison to single-beam (●) values and linearity disk (▲) values.

giving a change of $10:1$ in slitwidth. The double-beam errors at the new slitwidth are presented in Fig. 5. Following the large slitwidth change, the photometric error was still 0.2% or less above the 10% T level. Below the 10% T level, the error became progressively more severe, approaching 0.7% near the instrument zero.

Conclusions—In view of the above results, we feel a system such as we have described can be of valuable use when high photometric accuracy is required in an optical null infrared spectrophotometer. Double-beam accuracies of better than 0.2% have been demonstrated throughout the entire %T range at fixed frequencies. Equally high accuracy can be expected over frequency ranges where slitwidth changes are not extremely large. Over wide frequency ranges (large slitwidth changes), double-beam accuracies of better than 0.2% have been obtained, except at very low %T values.

Technique for Establishing Baseline in Infrared Spectrometry

2.24

R. O. Crisler and I. M. Brubaker*

The Procter & Gamble Company, Ivorydale Technical Center, Cincinnati, Ohio 45217

The determination of the shape and intensity of an infrared absorption band is complicated by the difficulty of establishing the zero absorbance (100% transmittance) baseline to which the absorption band may be related. For example, the choice between a Gaussian or Cauchy distribution function as a model for the absorption band is particularly affected because these functions differ most significantly in ordinate at distances greater than about one-half-bandwidth from the band center, where the relative error in transmittance will also be the largest.[1] The use of blanks for determining the zero-absorbance line (either empty cells or cells filled with pure solvent) is of limited value, particularly in infrared spectroscopy, because of the variations in refractive index of the solution at different concentrations of solute and the resultant variations in reflection losses at the cell–solution interfaces.

A technique using a thin cell and a variable space cell at two thicknesses was recently suggested by White and Ward[2] for eliminating interference fringes and compensating for dirt and fog on the cell windows. We wish to point out that a variation of this technique, in which spectra of a solution are obtained

---

*Present address: Universal Oil Products Co., Des Plaines, Ill.

1. R. N. Jones, K. S. Seshadri, N. B. W. Jonathan, and J. W. Hopkins, Can. J. Chem. **41**, 750 (1963).

2. J. U. White and W. W. Ward, Anal. Chem. **37**, 268 (1965).

at two thicknesses in a variable space cell, may be used to determine a correct zero-absorbance line.

If the various cell and reflection losses are represented by the factor R, the absorbance law becomes

$$P/P_0 = R \exp(-abc),$$

where P/P_0 is the intensity ratio of transmitted to incident radiation passing through a layer of thickness b, of a solution of material having a concentration c and an absorptivity a. If the spectrum is obtained at two thicknesses in a variable space cell, assuming that the unwanted losses are the same for the cell windows and solutions, then the relation between the ratio of intensities and the two thicknesses is

$$P_2/P_1 = \exp[-ac(b_2 - b_1)],$$

or, in absorbance units,

$$A_2 - A_1 = ac(b_2 - b_1).$$

This relation should be true for systems which obey Bouguer's law.[3]

To illustrate this technique, infrared spectra of cyclohexane were obtained at two thicknesses in a variable space cell without removing the cell from the instrument. A baseline was also obtained with no cells in the instrument. Spectra were obtained on a Perkin–Elmer model 521 spectrophotometer equipped with a DDR–1 digitizer. The spectra, shown in Fig. 1, were digitized at 0.2-cm^{-1} intervals, processed and plotted by computer.

The absorbance difference between the "thick" and "thin" spectra of Fig. 1 and a synthetic spectrum calculated as the sum of two Cauchy functions are plotted in Fig. 2. The Cauchy functions are centered at 903.28 and 861.26 cm^{-1} and have half-widths of 5.82 and 9.50 cm^{-1}. The difference between the ob-

3. R. P. Bauman, *Absorption Spectroscopy* (John Wiley & Sons, Inc., New York, 1962).

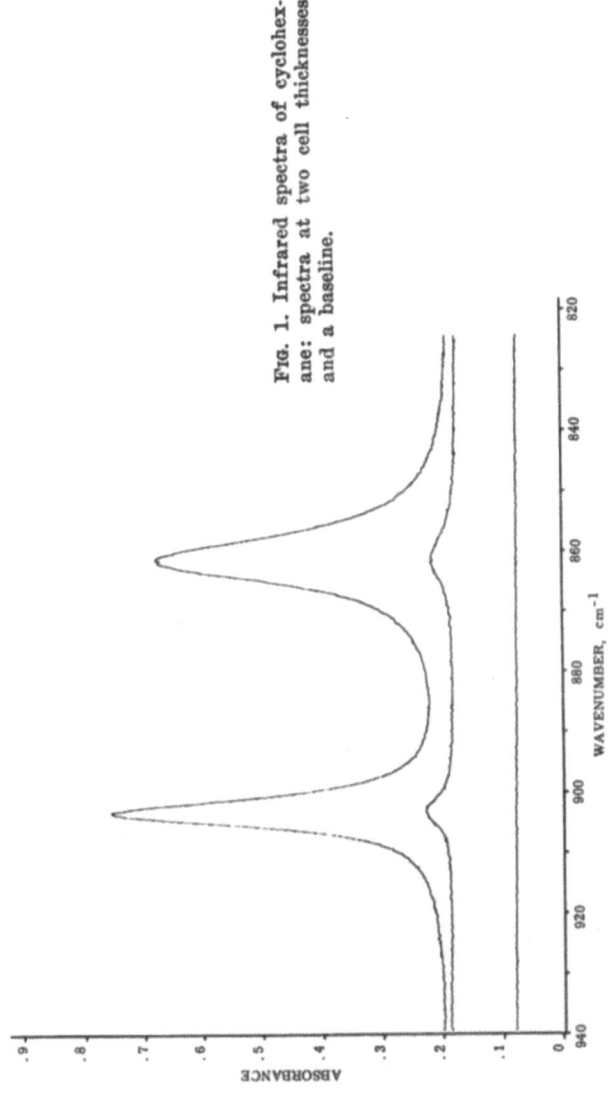

Fig. 1. Infrared spectra of cyclohexane; spectra at two cell thicknesses and a baseline.

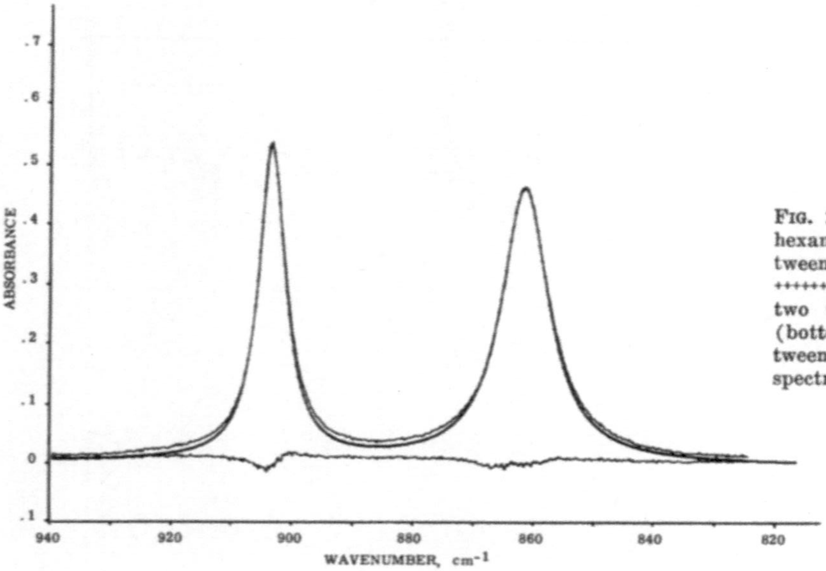

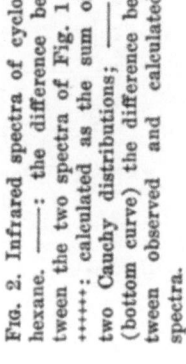

FIG. 2. Infrared spectra of cyclohexane. ———: the difference between the two spectra of Fig. 1;
+++++: calculated as the sum of
two Cauchy distributions; ———:
(bottom curve) the difference between observed and calculated
spectra.

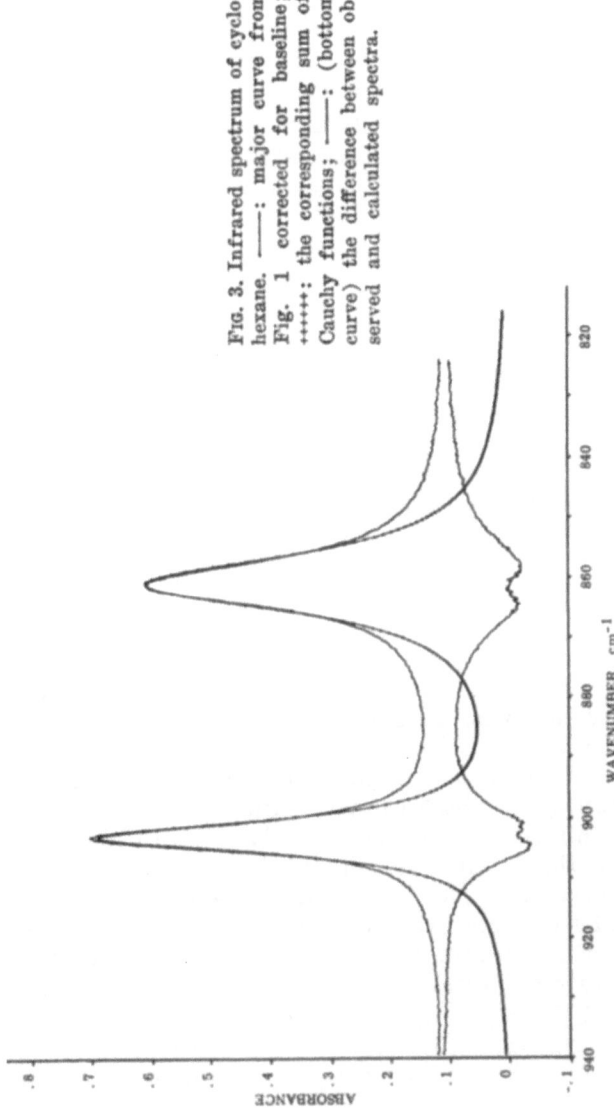

Fig. 3. Infrared spectrum of cyclohexane. ———: major curve from Fig. 1 corrected for baseline; ++++++: the corresponding sum of Cauchy functions; ———: (bottom curve) the difference between observed and calculated spectra.

served and calculated spectra is also shown. For comparison, the "thick" spectrum from Fig. 1, corrected for instrument baseline, and the attempt to fit it with Cauchy distributions is shown in Fig. 3.

We have successfully used this technique for the study of instrument spectral slit functions and dynamic response as well as for the study of infrared band contours.

2.25 A Nomograph for the Correction of Spectral Data Recorded in Transmittance

J. H. Gould and J. T. Chen

Division of Food Chemistry and Technology, Bureau of Science, Food and Drug Administration, Department of Health, Education, and Welfare, Washington, D. C. 20204

Many spectra, particularly infrared spectra, are recorded on charts which are linear in transmittance as the ordinate vs wavelength or wave number as the abscissa. When these records are used for quantitative analysis, the data must be converted to absorbance.

The first step in this conversion consists of correcting the recorded transmittance at the point of minimum transmission of the band for the departure of the baseline[1] or of the background[2] reference curve—the I_o level—from the 100% line of the chart. This correction can be quite large when the measurements are made using a single beam instrument; in this case, recordings made even under the best con-

1. W. J. Potts, Jr., *Chemical Infrared Spectroscopy* (John Wiley & Sons, Inc., New York, 1963), Vol. 1, p. 165.

2. R. P. Bauman, *Absorption Spectroscopy* (John Wiley & Sons, Inc., New York, 1962), p. 372.

ditions of sample preparation will result in the I_o level approximating a black-body curve.

In dealing with a spectrum recorded by a double-beam instrument, the reference level is generally much more nearly flat, and for a true solution of a sample with well-separated bands, the instrumental conditions can usually be adjusted so that the I_o level will approximate the 100% line of the chart. This condition will not exist, however, if the anɩ- lytical band is associated closely with overlapping neighboring bands.[3,4]

A departure from the flat baseline condition will also occur when the sample scatters the radiation passing through it. This condition is most frequently observed when a solid sample is not completely dispersed, as in a mull or alkali halide pellet. Some samples consistently resist all efforts to disperse them sufficiently, and the mathematical correction of the band transmittance must be made.

The fact that real experimental difficulties are often associated with attaining a coincidence of the nonabsorbing regions of the spectrum with the 100% line of the chart has been incorporated into the design of the correction nomograph, but it forces the requirement that the instrumental zero be made coincident with the chart zero, either by use of an opaque shutter or, when stray radiation is present, a thick sample.[5] This arises because the equation of the conversion nomograph—called on "N" or "Z" chart[6]—is applicable to the solution of an equation of the type

$$f_1(u) = f_3(w)/f_2(v),$$

3. See Ref. 1, p. 168.

4. A. L. Smith, Appl. Spectry. **12**, 153 (1958).

5. W. J. Potts, Jr., *Chemical Infrared Spectroscopy* (John Wiley & Sons, Inc., New York, 1963), Vol. 1, p. 162.

6. A. B. Calder and D. A. Calder, Roy. Inst. Chem. (London), Lectures, Monographs, Rep. **1962**, 21 (1962).

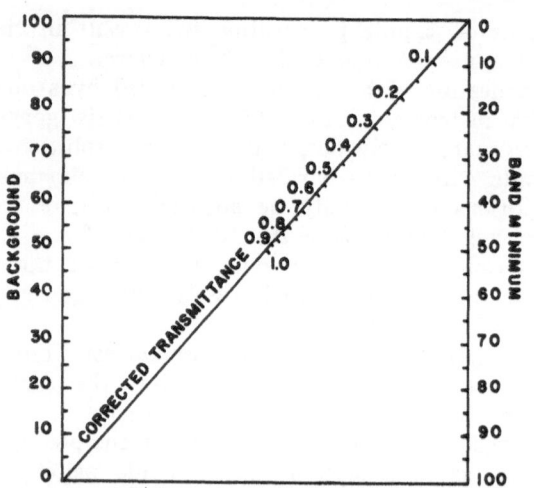

FIG. 1. A nomograph for converting recorded transmission to true transmittance.

where, in this case, $f_1(u)$ is the corrected transmittance, $f_2(v)$ is the scale readings of I_o, and $f_3(w)$ is the scale value of the transmission at the band minimum. Such an equation obviously has no capacity for treating the additional variable which would be introduced if deviations were allowed between the chart and instrumental zeros.

This nomograph is constructed as shown in Fig. 1 by drawing two parallel lines of any convenient length and distance apart on rectilinear graph paper. Although these lines can be arranged in many ways with respect to the amount of overlap one has relative to the other, the best use is made of the available space if they are arranged as shown. Each line is scaled to contain one hundred units, the left line with zero at the bottom, the right line with zero at the top.

The third scale is located on the diagonal line connecting the zero marks of the right and left scales.

This diagonal scale is calibrated by first placing a straight edge so that it connects the two 100 unit points of the side scales and marking its intersection with the diagonal by a "tick" which is given the value of 1.0 (100%). Additional marks are placed on the diagonal scale by keeping the straight edge centered on the left scale value of 100 while it is moved to intersect the right scale at 90, 80, 70, etc., in turn. The successive intersections with the diagonal line are marked and given the respective values of 0.9, 0.8, 0.7, etc. The subdivisions of these intervals are added to the diagonal in the same way.

The transmittance correction is carried out by (1) locating the I_o value, as read from the spectral record, on the left scale of Chart 1; (2) locating the scale value of the minimum reading of the band being corrected on the right scale; and (3) connecting these two points with straight edge. The corrected transmittance is read from the intersection of the straight edge and the diagonal line.

The corrected transmittance is then converted to absorbance by the use of a table of logarithms of decimal fractions, or by especially prepared conversion tables.[7,8]

A graphical method of conversion consists of drawing a diagonal line on a sheet of single cycle, semilogarithmic graph paper which connects the value 10 (given a value of 1.0 for this application) on the log axis with a convenient point on the linear axis which is some suitable multiple of ten from the origin.

The log scale of the chart is labeled transmittance and directly covers the analytically important range of 0.1–1.0 units. The linear scale is labeled absorbance and is given values of zero at the origin and 1.0

7. W. R. Brode, *Chemical Spectroscopy* (John Wiley & Sons, Inc., New York, 1943), 2nd ed., p. 615.

8. D. H. Simmonds, Anal. Chem. **30**, 1043 (1958).

at the intersection of the diagonal with the linear axis.

When the corrected value so found is in the range of values from 1.0 (100%)–0.1 (10%), that value is located on the transmittance axis, followed across horizontally to the intersection with the diagonal line, then vertically down to the absorbance axis, where the converted value is read directly in absorbance.

2.26   Direct Determination of the Integrated Absorption Intensity from an Infrared Transmission Curve

P. A. H. M. Hol

Central Laboratory, Staatsmijnen in Limburg, Geleen, Netherlands

A method is described that makes it possible to determine the area of an absorption band from the

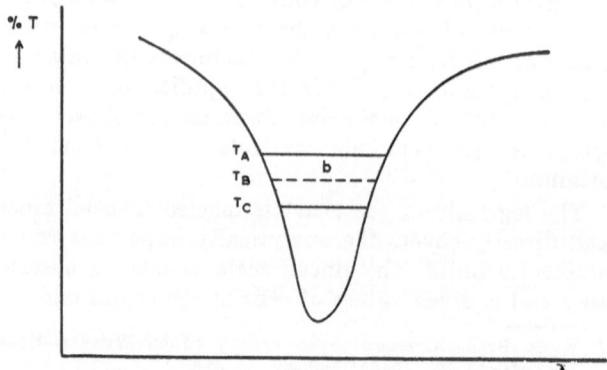

FIG. 1. Shape of an infrared transmission band.

transmission band without previously converting the percentage transmission to extinction values.

Suppose that the band has the shape illustrated in Fig. 1. The area enclosed by the curve can be divided into a number of trapezoids. The area of such a trapezoid in the corresponding absorption curve may be expressed as $S_B = b \log T_A/T_C$, where b is the half-width of the trapezoid and T_A and T_C are transmission values.

Provided that the interval between T_A and T_C is small (e.g., 1%), the half-width of the chosen trapezoid is the same in the transmission and in the absorption curve. When $T_A - T_C = 1\%$, T_A may be replaced by $T_B + \frac{1}{2}$ and T_C by $T_B - \frac{1}{2}$ because, at small intervals between T_A and T_C, $T_A T_C \simeq T_B^2$. Hence,

$$S_B = b \log (T_B + \tfrac{1}{2}/T_B - \tfrac{1}{2}),$$

or

$$S_B = b \log e \ln\frac{T_B + \frac{1}{2}}{T_B - \frac{1}{2}} \tag{1}$$

$$= b \log e \{\ln(2T_B + 1) - \ln(2T_B - 1)\}$$

$$= b \log e \left(\frac{1}{T_B} + \frac{1}{12T_B^3} + \frac{1}{80T_B^5} + \cdots \right).$$

In general, the value of T_B ranges between 30 and 100; so, terms with powers of T_B higher than 1 may be neglected. Hence,

$$S_B = (b/T_b) \log e. \tag{2}$$

The total area enclosed by the curve is then

$$S_T = \log e \Sigma (b/T_b). \tag{3}$$

What is the best way to determine b/T_b? We answer this question with reference to a transmission curve recorded with a Perkin–Elmer model 21 in-

strument. The ordinate on the recording paper is a
transmission scale of 200 mm length.

A flat Perspex plate is divided vertically into percent of transmission in such a way that the division
corresponds exactly to the scale on the recording
paper. Perpendicular to this scale, a straight line,
divided into millimeters, is drawn that intersects the
transmission scale at 100%. A revolving pointer,
provided with a black groove, is fixed in the origin.
The procedure is outlined in Fig. 2. Suppose that
we want to determine the size of the area between
60.5% and 59.5% transmission. The spectrum is covered with the Perspex plate in such a way that the
60% point coincides with the 60% transmission point
on the left in the spectrum. The groove in the pointer
is placed over the 60% point in the right-hand part
of the curve. The value a, read from the millimeter

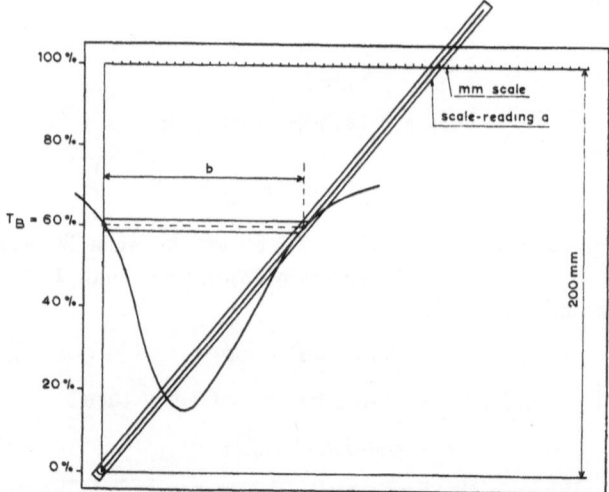

F<small>IG</small>. 2. Procedure for measuring the area enclosed by a transmission curve.

scale, is now a direct measure of the unknown area, since $b/T_b = a/100$. To find the total area enclosed by the curve, we summarize the scale readings, obtained by placing the transmission scale at 59%, 58%, etc. The time required for the summation is about $\frac{1}{4}$ h. In cases where the half-width of the band is small, the abscissa can be expanded to make the calculation more accurate.

Especially when it is desired to know the relative values of the integrated intensity, the technique described is very convenient. The usefulness of the method is illustrated with reference to the determination of the ratio of the integrated absorption intensities at 965 cm^{-1} in (a) 17-pentatriacontene, mixed with n-hexatriacontane and in (b) 17-pentatriacontene, dissolved in cyclohexane. In general, it is to be expected that the ratio of these intensities equals one. The results are summarized in Table I.

The deviation of the ratio from 1 is negligible. The agreement between the planimetric value and the value obtained by the proposed method is excellent. *Acknowledgment.* The author is indebted to Dr. R. J. de Kock for the critical reading of the manuscript.

Applied Method	Ratio $A_{\text{SOLID}}/A_{\text{LIQUID}}$	
Lorentz curve[a] $A = \dfrac{\pi}{2cl} \ln\left(\dfrac{I_0}{I}\right)_{\text{max}} \Delta\nu^{\frac{1}{2}}$	1.13	Table I. Ratio of the integrated absorption intensities of the CH-out-of-plane bending absorption in the solid and the liquid states of aggregation (A_S/A_L). c is the concentration (mole/L). l is thickness in cm.
Planimetry area/cl	1.02	
Proposed method area/cl	1.03	

[a]C. L. Lau, Spectrochim. Acta **14**, 181 (1959); D. A. Ramsay, J. Am. Chem. Soc. **74**, 72 (1952); and R. N. Jones *et al.*, *ibid.* **74**, 80 (1952).

2.27   Numerical Method for the Direct Determination of Integrated Absorption Intensities from Transmission Curves

A. L. Khidir Aljibury

College of Science, University of Baghdad, Baghdad, Iraq

In the determination of integrated intensities, one usually replots the transmission curve point by point on a logarithmic scale, then integrates between practical frequency limits to obtain the integrated intensity

$$\int_{\nu_B}^{\nu_A} \ln \frac{I_0}{I} \, d\nu. \tag{1}$$

The geometrical method suggested by Hol[1] requires the transmission curve and a geometrical device to determine the integrated intensity. In this note a numerical method is outlined for the determination of integrated intensities which requires the transmission curve and a logarithm table.

Method. The integral [Eq. (1)] may be broken down into the sum of areas of small trapezoids in the following manner:

$$\int_{\nu_B}^{\nu_A} \ln \frac{I_0}{I} \, d\nu = \sum_{m=100}^{m=T_{min}} b_{m,m-1} \ln \frac{m}{m-1}, \tag{2}$$

i.e., the height of the trapezoid (interval) is 1%. $b_{m,m-1}$ is the half-width of the trapezoid between $\ln (I_0/m)$ and $\ln (I_0/m-1)$.

If the interval is small, such that the line segments of the logarithmic curve and of the transmission curve are approximately straight lines, then by simple geom-

---

1. P. A. H. M. Hol, Appl. Spectry. **20**, 52 (1966).

Table I. A logarithm table for use in conjunction with transmission curves for the calculation of integrated absorption intensities.

m	$\ln (m/m-1)$	m	$\ln (m/m-1)$
100	0.0100	92	0.0108
99	0.0101	91	0.0109
98	0.0102	90	0.0110
97	0.0103	89	0.0112
96	0.0104		
95	0.0105		
94	0.0106		
93	0.0107	etc.	

etry $b_{m,m-1}$ of the trapezoid is more or less the same as the half-width of the corresponding trapezoid between m and $m-1\%$ transmissions. Therefore,

$$\sum_{m=100}^{m=T_{\min}} b_{m,m-1} \ln \frac{m}{m-1} \qquad (3)$$

can be calculated by accumulative multiplication, using a desk calculating machine, of the appropriate

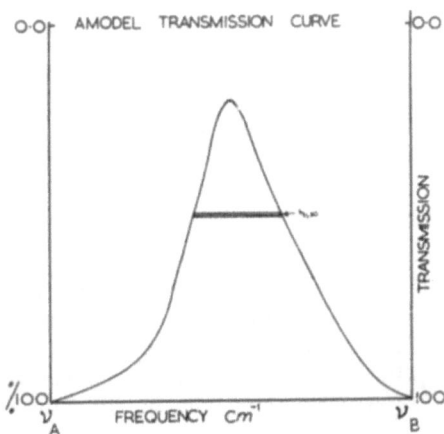

Fig. 1. A model transmission curve.

half-width $b_{m,m-1}$, which can be read directly from the transmission curves, and the corresponding value of $\ln(m/m-1)$ from the standard table of $\ln(m/m-1)$ (see Table I).

As a model calculation the integrated intensity of the curve shown in Fig. (1) is evaluated. The time consumed is entirely that which one can read off the b values from the curve. The value obtained for the integrated intensity by the numerical method is 82.2 cm^{-1}. The integrated intensity found by measuring the area under the curve of $\ln(I_0/I)$ was found to be 81.2 cm^{-1} (using a planimeter). The author believes that this method is simple and faster than that suggested by Hol if not more accurate, since the latter method requires the following approximation

$$1/T + C/T^2 + \cdots \text{etc.} \simeq 1/T \tag{4}$$

which the present method does not.

The author is grateful to Dr. G. A. W. Derwish and M. F. Mayahi for useful consultation.

2.28 Measurement of the Apparent Bandwidth Parameters of an Absorption Band

A. M. Kabiel and C. H. Boutros

Medical Research Institute, Physics Department, Alexandria, U.A.R.

It is quite well known that an absorption spectrum, either recorded or point–point measured, is usually represented graphically as a percentage transmission or absorption vs wavenumber or wavelength.[1] For such an absorption band, the baseline[1] is carefully

1. W. Brugel, *An Introduction to Infrared Spectroscopy* (Methuen and Company Ltd., London, 1962).

selected. Usually, it is tangential to the band wings and the different values of T_0 and T (where T_0 and T are the apparent incident and transmitted intensities corrected for background at different wavenumbers or wavelengths) are thus measured from it. Hence, all absorbance-wavenumber or wavelength-derived bands can be illustrated graphically, from which the different apparent bandwidth parameters $\Delta\nu_{\frac{1}{2}}^a$, $\Delta\nu_{\frac{3}{4}}^a$, and $\Delta\nu_{\frac{1}{4}}^a$ measured at $\frac{1}{2}$, $\frac{3}{4}$, and $\frac{1}{4}$ the maximum absorption can be directly estimated.

The aim of this note is to introduce a new method to estimate the different apparent bandwidth parameters directly from the experimental absorption band.

Experimental measurements were carried out on the CO-stretching vibration at 1682 cm^{-1} of acetophenone in chloroform, as illustration. Different concentrations were prepared and the full absorption band was recorded for one slit program[2,3] using a UR-10 infrared spectrophotometer. Fig. 1 shows such an absorption band for a selected value of $(cl) = 0.1305 \times 10^{-2}$ Mole·litre^{-1}·cm, where c denotes the concentration and l the cell thickness.

Let T_{0m} and T_m denote the apparent incident and transmitted intensities corrected for background at maximum absorption of the selected band, and let T_0 and $T_{\frac{1}{2}}$ denote the apparent incident and transmitted intensities corrected for background at the position where the $\Delta\nu_{\frac{1}{2}}^a$ parameter is to be measured. Thus,

$$\tfrac{1}{2} \log (T_0/T)_m = \log (T_0/T_{\frac{1}{2}})$$

or

$$T_{\frac{1}{2}} = \left[T_m (T_0^2/T_{0m}) \right]^{\frac{1}{2}}. \tag{1}$$

Assuming that the absorption band under investigation is isolated, then T_{0m} is very nearly equal to T_0, and thus Eq. (1) will have the form

2. A. M. Kabiel *et al.*, Z. Physik. Chem. 228, (1965).
3. A. M. Kabiel *et al.*, Z. Physik. Chem. 234, (1967).

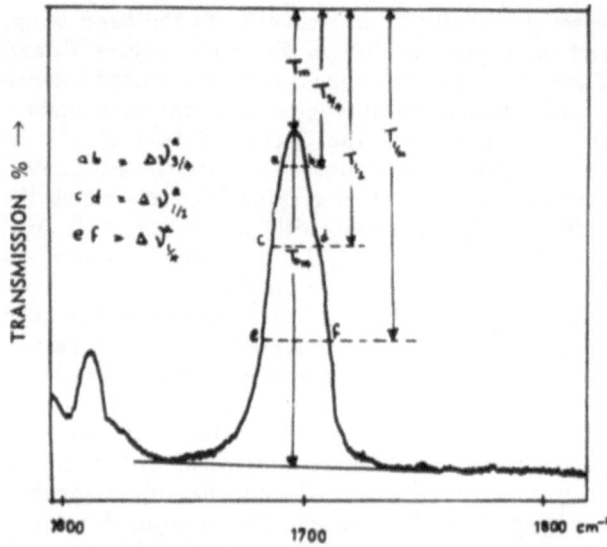

FIG. 1. Variation of percentage transmission vs wavenumber
of the CO–stretching vibration at 1682 cm of acetophenone in
chloroform (cl=0.1305×10⁻² mole·liter⁻¹·cm).

$$T_{\frac{1}{2}} = (T_m \cdot T_{0m})^{\frac{1}{2}}. \qquad (2)$$

On the other hand, if $T_{\frac{3}{4}}$ and $T_{\frac{1}{4}}$ are the transmitted
intensities corrected for background at which the $\Delta \nu_{\frac{3}{4}}^{a}$
and $\Delta \nu_{\frac{1}{4}}^{a}$ parameters are measured, respectively, then

$$T_{\frac{3}{4}} = (T_m{}^3 \cdot T_{0m})^{\frac{1}{4}} \qquad (3)$$

and

$$T_{\frac{1}{4}} = (T_m \cdot T_{0m}{}^3)^{\frac{1}{4}}. \qquad (4)$$

If the band was not isolated, T_0 had to be measured on
both sides of the maximum absorption. The arithmetic
mean of the 2 measurements will be very nearly equal
to T_{0m} so that Eqs. (2), (3), and (4) can be applied.
Moreover, for such a band, $\Delta \nu_{\frac{3}{4}}^{a}$ will be the more representative parameter to obtain better results than

Table I. Bandwidth parameters estimated using both absorbance and proposed method for acetophenone in chloroform. Band at 1682 cm^{-1} (cl=0.1305×10^{-5}).

log (T$_0$/T)$_m$	$\Delta\nu_{\frac{1}{4}}^a$		$\Delta\frac{1}{2}\nu^a$		$\Delta\nu_{\frac{3}{4}}^a$	
	Absorb-ance curve	Absorp-tion curve	Absorb-ance curve	Absorp-tion curve	Absorb-ance curve	Absorp-tion curve
0.900	18.3	18.2	11.2	11.1	26.3	27.9
0.706	18.5	18.6	11.3	11.2	28.7	28.8
0.574	18.8	19.0	11.5	11.4	29.0	28.6
0.311	18.9	19.0	11.5	11.3	29.2	29.0
0.193	19.0	19.2	11.5	11.6	29.5	29.4

$\Delta\nu_{\frac{3}{4}}^a$. $\Delta\nu_{\frac{1}{4}}^a$ will not be, in this case, applicable because of the increased deviation at the wings.

Results obtained (Table I) using the proposed method to estimate both $\Delta\nu_{\frac{1}{4}}^a$ and $\Delta\nu_{\frac{1}{2}}^a$ are quite in agreement with those obtained from absorbance curves. The error is within 1%. In case of $\Delta\nu_{\frac{3}{4}}^a$ parameter, the error is slightly increased but, remains within 1.5%. This is attributed to the lack of perfect symmetry at the band wings.

It appears that the main advantage of the proposed method is to spare time and detailed routine measurements since it produces results adequate in approach to accuracy and precision of those obtained from the adsorption curves.

2.29 Convenient Method for Encoding Infrared Spectra Linear in Wavenumber Using the Sadtler Notation

Arthur Kramer

U. S. Customs Laboratory, New York, New York 10014

The Sadtler system is well suited for encoding infrared spectra that are displayed on a linear wavelength scale. However, for spectra presented on a scale linear in wavenumber, the conversion of the wavenumber values of the strongest peaks within each 1 μ interval into microns is tedious. The use of a transparent overlay for conversion requires maintenance of exact alignment of paper and overlay and suffers from the fact that the tenths of a micron lines in the the 13- and 14-μ region are closely spaced and the spacings vary in a nonlinear manner, thereby rendering visual interpolation difficult.

At the New York Customs Laboratory, we have devised a system to expedite the use of the Sadtler system for encoding spectra that are linear in wavenumber. Our method consists essentially of the use of a table of critical wavenumber values. These are defined as those wavenumber values equivalent to points midway between each tenth of a micron line on a linear wavelength scale.

The table is drawn up for the 9-μ region, for example, by successively dividing into 10 000 the numbers 9.05, 9.15, 9.25 $\cdots$ 9.95. The quotients are the critical wavenumbers and they are arranged in a column headed by an italic nine. The difference between any two successive critical wavenumbers in a column represents 0.1 μ. The tenths of a micron values are placed in the table between their corresponding wavenumber pairs and slightly to the left, as shown

Table I. Critical wavenumbers.

	0	3	4	5	6	7	8	9	10	11	12	13	14
0	4878	3279	2469	1980	1653	1418	1242	1105	995	905	830	766	712
1	4651	3175	2410	1942	1626	1399	1227	1093	985	897	823	760	707
2	4444	3077	2353	1905	1600	1379	1212	1081	976	889	816	755	702
3	4255	2985	2299	1869	1575	1361	1198	1070	966	881	810	749	697
4	4082	2899	2247	1835	1550	1342	1183	1058	957	873	803	743	692
5	3922	2817	2198	1802	1527	1325	1170	1047	948	866	797	738	687
6	3774	2740	2151	1770	1504	1307	1156	1036	939	858	791	733	683
7	3636	2667	2105	1739	1481	1290	1143	1026	930	851	784	727	678
8	3509	2597	2062	1709	1460	1274	1130	1015	922	844	778	722	673
9	3390	2532	2020	1681	1439	1258	1117	1005	913	837	772	717	669

in Table I. The width of the constructed table is approximately that of the chart. The variable spacing between each column is chosen so that every column lies in the region of the chart where its values are found when the chart and table are roughly aligned laterally.

In utilizing the table, it is placed beneath the spectrum, roughly aligned laterally, and the wavenumber of the strongest peak in each integral micron interval is used to enter the table. The eyes need only scan the column directly beneath the peak rather than the entire table. If the peak should chance to coincide with a critical wavenumber, the tenths of a micron value utilized for the code is the one located immediately below and to the left of the critical value.

Only a typewriter and a calculator or table of reciprocals are needed to quickly draw up the table. We have found that the use of the table significantly shortens the time and effort required for encoding and eliminates the problem of interpolating from a nonlinear scale.

2.30   Card Coding of Spectroscopic Data

D. S. Erley

Dow Chemical Co., Midland, Michigan

The proliferation of small computers and time-sharing terminals has made mechanical card sorting systems virtually obsolete for spectroscopic information retrieval. The card format restrictions imposed by mechanical systems make the data cards difficult to process in many computers and require "throwing away" a great deal of valuable data in the coding process. Laboratories planning to code and

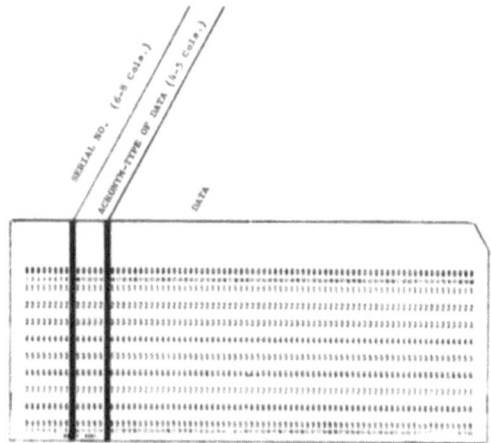

F<small>IG</small>. 1. General data-card format.

store spectroscopic data on cards should recognize
that these cards will be read only to generate secondary tape or disk files for computer processing. Therefore, the one-card-per-compound format should be
abandoned and a more flexible one such as that shown
in Fig. 1 adopted.

Each card should contain the serial number of the
compound, an acronym for the type of data stored,
and the data itself. All the data for one compound
may then be stored as a consecutive series of cards.
If more than one card is needed for a particular type
of data, e.g., absorption band positions, it is necessary only to reproduce the serial number and acronym fields on each. The acronym should also provide
a key to the format of the data.

When coding data, exact chart readings should be
recorded wherever possible. Many of these data may
be ignored by the computer when a working file is
generated, but the necessity for recoding the file at
a later time often can be avoided.

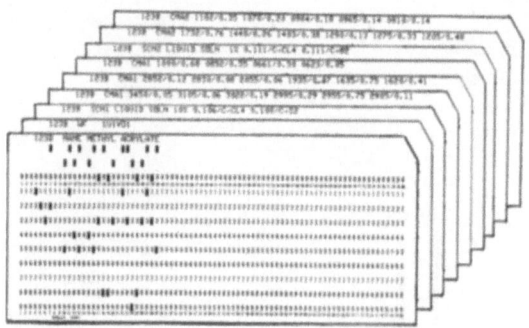

FIG. 2. Cards coded for typical infrared standard including: Name, Wiswesser notation, Scan parameters (state, concentration, cell length, solvents), Absorption band positions/absorbances.

Structural information is probably best coded in Wiswesser line formula notation.[1,2] The "Wiswesser formula" gives the complete molecular structure of a compound and can, therefore, be used for fragment searching. It usually requires fewer characters than the name of the compound.

Figure 2 shows the cards coded for a typical infrared spectrum. Each may be identified by its acronym. A fixed format is used for each type of data to facilitate translation.

Acronym:	*Meaning:*
NAME	Name of compound.
WF	Molecular formula in Wiswesser Notation.
SCN1	Scan parameters for 10% scan.
CMA1	Absorption band positions (wavenumber/absorbance) in 10% scan.
SCN2	Scan parameters for 1% scan.
CMA2	Absorption band positions in 1% scan.

1. W. J. Wiswesser, J. Chem. Doc. 8, 146 (1968).
2. E. G. Smith, *The Wiswesser Line-Formula Chemical Notation* (McGraw–Hill Book Co., New York, 1968).

The format suggested herein may be used for many types of data. The rules are general enough to permit adapting it to individual laboratory needs. Nevertheless, the program changes needed to recognize most modifications are relatively minor, so the exchange of data between groups, or the programs for handling such data, will be greatly facilitated.

SECTION 3
MASS SPECTROSCOPY

New Method for the Production of Ions   **3.1**
from Nonconductors for Analysis by Solid-
Source Mass Spectrography

P. Deines, T. J. Eskew, and L. F. Herzog

*The Pennsylvania State University, Dept. of Geology and
Geophysics, University Park, Penna. 16802*

In recent years, the mass spectrograph has become
an important tool for the chemical analysis of conductors, semiconductors, and nonconductors. Dempster[1] first applied the radiofrequency (rf) spark to
the analysis of nonconductors during the Manhattan
Project of the early 1940's as a means of prospecting
for uranium and its isotopes in minerals; and in the
early 1950's, Hannay and Ahearn[2] extended the technique to the analysis of semiconductors. The vacuum-
vibrator source (also referred to as the dc vibrator
arc) was also first used by Dempster,[3,4] in the 1930's.
Since 1950, its development into an analytical tool for

1. A. J. Dempster, MDDC Report **370**, U. S. Dept. Comm.,
   Office Tech. Serv. P. B. Rept. [also reported by M. G.
   Inghram in Tech. Publ. **149**, ASTM Std., (1953).]
2. A. J. Ahearn and N. B. Hannay, Anal Chem. **26**, 1056–
   1058 (1954).
3. A. J. Dempster, Nature **135**, 542 (1935).
4. A. J. Dempster, Rev. Sci. Instr. **7**, 46–49 (1936).

conductors has been carried forward through the work of Preston,[5] Herzog,[6] Venkatasubramanian,[7] and Schuy,[8,9] among others. To the best knowledge of the authors, the vibrator has not been successfully used for the ionization and analysis of nonconducting materials until now.

The vacuum-vibrator source operates at low voltages (under 100 V compared to the 40 000 to 100·000 V used in an rf-spark source), so that the main problem in applying it to nonconductors is that of devising a means of initiating and sustaining a discharge between nonconducting sample electrodes, or, alternatively, between one such electrode and a metal electrode.

We wish to report the development of a technique whereby the ionization of nonconductors such as Al_2O_3, MgO, SiO_2, and rocks and minerals can be accomplished with a vacuum-vibrator source with good efficiency. The sample is ground to a fine powder (200 mesh) and mixed with precipitated silver powder. As examples, 20% MgO or Al_2O_3, by weight, mixed with 80% silver, yield atomic ratios of approximately 0.67 for Mg/Ag and 0.53 for Al/Ag. The silver-matrix powder and the sample powder are thoroughly mixed, ground together, and pressed under 100 000 psi into electrode rods 0.1 in. in diam and 0.15 to 0.2 in. long. Such electrodes possess sufficient rigidity to withstand the mechanical vibrations to which they are subjected in the vacuum-vibrator (VV) ion source that we used.

5. R. S. Preston, M. S. thesis (Wesleyan Univ., Middletown, Conn., 1950).
6. L. F. Herzog, *Symposium on the Analysis of Solids by Mass Spectroscopy*, at the International Conference on Spectroscopy, College Park, Md. (1962).
7. V. S. Venkatasubramanian, Can. J. Phys., **41**, 234–239 (1963).
8. K. D. Schuy, Z. Naturforsch **18a**, 95–97 (1963).
9. K. D. Schuy, Z. Naturforsch. **18a**, 926–941 (1963).

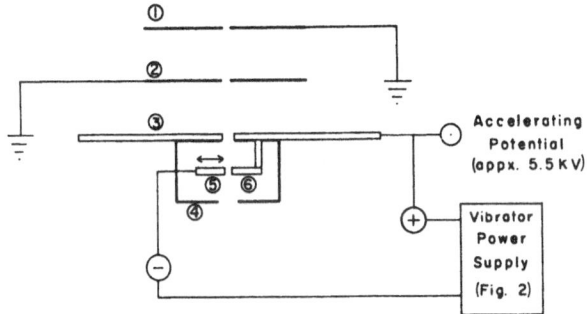

FIG. 1. Vibrator source: physical configuration. (1) and (2) are grounded ion-beam collimating plates; (3) is the source support plate, which is set at a suitable ion-accelerating voltage; (4) is a removable shield that collects sputtered material [it extends over the surface of (3) also]; (5) is the vibrating electrode; and (6); is the stationary electrode of the vacuum vibrator.

The instrument used in this study was a second-order double-focusing mass spectrograph having electrostatic deflection $\phi_e = 53.03°$, magnetic deflection $\phi_m = 90°$, electrostatic radius $r_e = 15.98$ cm, and magnet entrance boundary radius 6.67 cm; this geometry was proposed by Hintenberger and Konig[10] and built by us for ultratrace element studies.[11] For these experiments, the rf-spark source usually used was replaced by an experimental vibrator source developed at Nuclide Corporation.[12]

Operating conditions of the vibrator source were: dc vibrator voltage—70 V, accelerating potential—5.5 kV, limiting resistor—50 Ω, exposure time—5

10. H. Hintenberger, Advan. Mass Spectrometry, Proc. Conf. Univ. London, 34 (1959).

11. L. F. Herzog, T. J. Eskew, and P. Deines, Spec. Rept. No. −165 (Coll. Min. Ind., The Pa. State Univ., University Park, Pa., Dec. 1965), pp. 1–179.

12. L. A. Cambey, Tech. Documentary Rept. No. ML-TDR-64-75, (A. F. Materials Lab., Res Tech. Div., Wright–Patterson AFB, Ohio, 1964).

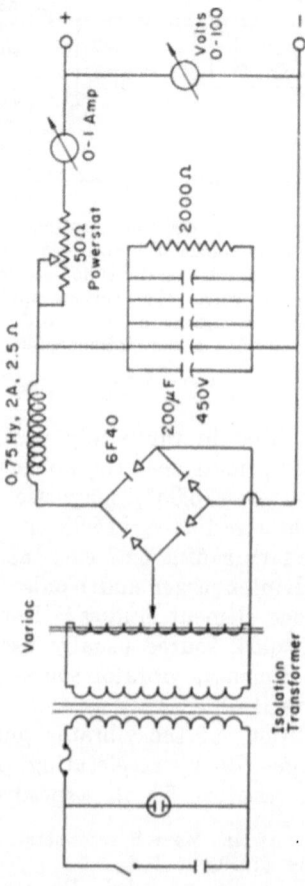

FIG. 2. Vibrator source electronic control circuit.

min. Figure 1 shows the electrode configuration used and Fig. 2 shows the simple circuit required to operate the source.

In the spectra obtained with the vacuum vibrator, the singly charged species lines of major isotopes of the major elements in the nonconducting samples first appeared at about the same exposure levels as the lines of the silver matrix for all of the insulator–silver mixtures thus far investigated. In the case of an igneous rock sample, the Canadian Association for Applied Spectroscopy's standard sample "Syenite Rock-1",[13] all of the major and minor constituents of the rock, down to the 1% level, were recorded in exposures less than 10 sec long. Average currents of 5×10^{-11} A were achieved with this "breadboard" unit, while maximum ion currents of 10^{-8}–10^{-9} A were obtained and could, we believe, be sustained in a source improved in details of mechanical construction. A photodensitometer trace of one run on S-1, covering the mass range 20–120, is given as Fig. 3. In this 5-min run, lines that seem to be attributable to isotopes of Co, Cu, Zn (mass 60–70 region), and other species present in concentrations below 0.01% by weight were recorded. Note also that Ag^{2+} (m/e $54\frac{1}{2}$, $53\frac{1}{2}$) and Ag^+ (m/e $36\frac{1}{3}$, $35\frac{1}{3}$), for example, are of approximately equal abundance on this plate, and from this and other runs it is believed that U^+ and U^{2+}, Th^+ and Th^{2+}, and the singly and doubly charged ions of other constituents are likewise each of approximately equal abundance. From these analyses it is concluded that 100-ppm (atomic) species can be detected in runs of a few minutes duration with the VV source. Table I lists the elements that are most probably responsible for the observed lines on this plate, giving the "mean" concentrations given by

13. L. A. Cambey and T. J. Hovick, Tech. Rept. AFML-TR-65-81, (A. F. Materials Laboratory, Wright-Patterson AFB, Ohio, March 1965).

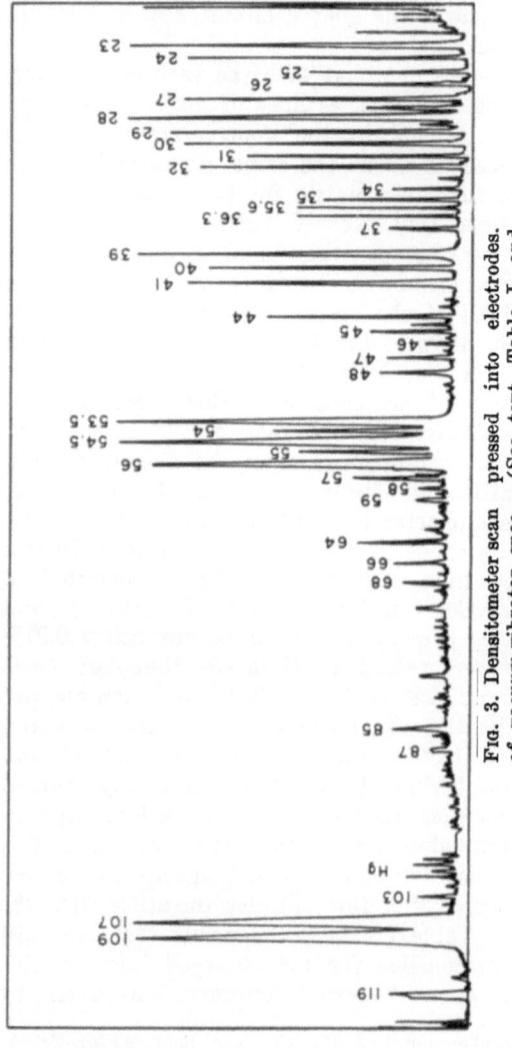

Fig. 3. Densitometer scan of vacuum-vibrator spectrum of powdered syenite rock sample S-1 (20% by weight) mixed with powdered silver and pressed into electrodes. (See text, Table I, and Fig. 4.) Note: densitometer scan rate is not constant.

FIG. 4. Print of *m/e* 20-to-120 portion of photoplate spectrum of syenite-rock–silver mixture (see Fig. 3).

Webber. Figure 4 is a reproduction of the plate (faint lines are lost in transfer to the printed image).

This plate, of course, gives only a very approximate indication of relative abundances because of the nonlinear response of photographic plates. One would expect, if there is no fractionation nor any matrix effect, that elements would rank in order of abundance on this plate, as they do in Webber's table. This is in general true, except as regards the Group I elements Na, K, and Rb, which are all too intense on the plate relative to the neighboring mass Group II elements Mg, Ca, and Sr. It is not presently known whether this is attributable to the presence of the thermionic-emission process of ionization (which would enhance Group I) or to contamination.

These experiments demonstrate that the vacuum-vibrator source can, in fact, be used for the analysis of nonconducting materials. The loss in sensitivity that results from the fact that the sample has to be diluted considerably with a conductor to give the electrode sufficient conductivity to initiate and maintain a vacuum discharge (and adequate mechanical rigidity to hold together as the electrodes vibrate) is small. For many types of work, this disadvantage is outweighed by several advantages the vacuum-vibrator source offers when compared to the rf-spark source, with which so great a dilution is generally not necessary. The advantages of the vibrator source include these:

• The amount of electronic circuitry required for the operation of the source is very small indeed (see Fig. 2).

14. G. R. Webber, Appl. Spectry. **15**, 159–161 (1961).

Table I. Elements possibly responsible for observed lines in Syenite Rock S–1 vibrator source spectra (Ag matrix), and weight-

m/e	Element	Reported concentrations (element × isotope)
23	Na^+	(2.2×1.00)
24	Mg^+	(2.4×0.79)
25	Mg^+	(2.4×0.10)
26	Mg^+	(2.4×0.11)
27	Al^+	(5.0×1.00), Fe^{2+}
27.5	Mn^{2+a}	
28	Si^+	(27.5×0.92), Fe^{2+}, N_2^+
29	Si^+	(27.5×0.05), N_2^+
30	Si^+	(27.5×0.03)
31	P^+	(0.08×1.00)
32	S^+	(0.07×0.95), Zn^{2+}, O_2^+
33	Zn^{2+}, S^+	(0.07×0.01)
34	Zn^{2+}, S^+	(0.07×0.04)
35	Cl^+	$(\ ? \ \times 0.75)^b$
35.6	Ag^{3+}	Matrix
36.3	Ag^{3+}	Matrix
37	Cl^+	$(\ ? \ \times 0.25)$
39	K^+	(2.2×0.93)
40	Ca^+	(2.0×0.97), Ar^+
41	K^+	(2.2×0.07)
42	Ca^+	
43	Ca^+	
44	Ca^+	(2.0×0.02), CO_2^+
45	Sc^+	(0.002×1.00)
46	Ti^+	(0.3×0.08)
47	Ti^+	(0.3×0.08)
48	Ti^+	(0.3×0.73)
49	Ti^+	(0.3×0.06)
50	Ti^+	(0.3×0.05)
53.5	Ag^{3+}	Matrix
54	Fe^+	(5.3×0.06)

a Ratios of charge states + and 2 + are not yet known, but ion concentrations are believed to be not greatly different.
b Concentrations not reported by Webber.
c Mercury diffusion pumps were in use.
d The isotopic composition of lead in this specimen is not known but doubtless is atypical because of the high relative concentrations of U and Th.

• The ions produced have a much smaller energy spread (a few tens of volts as compared to several thousand volts), hence the ion-utilization efficiency of a spectrograph with a vibrator source is greater than that of the same spectrograph equipped with an rf-spark source. The energy spread is in fact sufficiently low to make possible use of a direction-focusing mass spectrometer with a high ion-acceleration potential for experiments requiring only moderate resolution and abundance sensitivity.

concentrations given by Webber (1961), multiplied by relative
percent element abundance of isotope.

m/e	Element	Reported concentrations (element × isotope)
54.5	Ag^{2+}	Matrix
55	Mn^+	$(0.3 \quad × 1.00)$
56	Fe^+	$(5.3 \quad × 0.92)$
57	Fe^+	$(5.3 \quad × 0.02)$
58	Fe^+	$(5.3 \quad × 0.003)$ Ni $(0.005 × 0.68)$
59	Co^+	$(0.002 × 1.00)$
60	Ni^+	$(0.005 × 0.26)$
63	Cu^+	$(0.002 × 0.69)$
64	Zn^+	$(0.02 \quad × 0.49)$
65	Cu^+	$(0.002 × 0.31)$
66	Zn^+	$(0.02 \quad × 0.28)$
67	Zn^+	$(0.02 \quad × 0.04)$
68	Zn^+	$(0.02 \quad × 0.19)$
69.5	La^{2+}	$(0.03 \quad × 1.00)$
70	Ce^{2+}	$(0.1 \quad × 0.88)$
78	K_2^+	
85	Rb^+	$(0.03 \quad × 0.72)$
87	Rb^+	$(0.03 \quad × 0.28)$
88	Sr^+	$(0.04 \quad × 0.83)$
99	Hg^{2+}	$(\quad ? \quad × 0.10)$ pump[b,c]
99.5	Hg^{2+}	$(\quad ? \quad × 0.17)$ pump
100	Hg^{2+}	$(\quad ? \quad × 0.23)$ pump
100.5	Hg^{2+}	$(\quad ? \quad × 0.13)$ pump
101	Hg^{2+}	$(\quad ? \quad × 0.30)$ pump
102	Hg^{2+}	$(\quad ? \quad × 0.07)$ pump
103	Pb^{2+}	$(0.05 \quad × \quad ?)$[d]
103.5	Pb^{2+}	$(0.05 \quad × \quad ?)$
104	Pb^{2+}	$(0.05 \quad × \quad ?)$
107	Ag^+	$(Matrix × 0.51)$
109	Ag^+	$(Matrix × 0.49)$
116	Th^{2+}	$(0.13 \quad × 1.00)$
119	U^{2+}	$(0.24 \quad × 0.99)$

• Spectral lines obtained with a double-focusing ion
analyzer are sharper, and hence (a) a higher resolution can be achieved with a given analyzer, while at
the same time (b) greater sensitivity is achieved
with a photoplate detector (since a given number of
ions impinge on a smaller area of sensitive emulsion)
because of the smaller area of the line when the
vibrator is used.

• According to Schuy,[8] varying the accelerating potential does not cause large changes in the abundance
ratios of ions of low-mass elements.

• The quantity of gas produced in operating such a source is smaller.

• Use of electrical detectors and other sensitive circuits is much more readily achieved, since disturbing rf interferences are not present.

For these reasons, we feel the dc vacuum-vibrator source should become a valuable addition to the group of mass spectroscope ion sources useful for the study of nonconductors and semiconductors, as well as conductors.

Acknowledgments. This study was performed with an instrument built with support from the Air Force Cambridge Research Laboratory, the National Science Foundation, the Advanced Research Projects Agency, and this University. The prototype vacuum-vibrator source used was constructed under an Air Force contract from the Wright–Patterson Air Force Base to Nuclide Corporation. The authors are indebted to Dr. L. A. Cambey and T. J. Hovick of Nuclide for their helpful suggestions and cooperation in this work.

3.2 Improved Source Mounting for Thermal-Source Mass Spectrometer

David M. Creighton

Isochem Inc., Richland, Washington 99352 and Atlantic Richfield Hanford Co., Richland, Washington 99352*

In a thermal-emission source mass spectrometer the accelerating and beam-adjusting plates are gen-

*This work was performed under a contract between Isochem Inc. and the AEC for operation of the chemical processing plant. The present contractor to the AEC for operation of this plant is the Atlantic Richfield Hanford Co.

erally separated by glass or ceramic spacers which provide both geometric positioning and electrical insulation.

For the instrument in the author's laboratory (Nuclide Corporation, Type SU), two types of spacers are available, glass sleeves or spheres. In both cases, the entire assembly of plates is held together with glass or ceramic bolts and metal springs. Because of the special problems associated with the processing of radioactive samples, it has been found that the assembly with glass sleeves is more convenient to use than the spheres.

The plate assembly bolts have heretofore been supplied by the instrument manufacturer in several configurations of glass or ceramic with ends threaded for metal nuts or notched for retaining clips. A common fault of these bolts is the expense of manufacture and the notch sensitivity of the glass or ceramic. The bolts fracture easily sometimes throwing radioactively contaminated fragments some distance.

To overcome the breakage and to increase the ease of component assembly, a new bolt was designed. It consists of a pyrex rod flattened on one end to form a head and rounded on the other with shank dimensions of $1\frac{1}{8}$ in. $\times$ 0.190 in. The glass bolt is stress relieved and assembled with a spring[1] which is held in place by a compression-retaining ring.[2] (Fig. 1.) This ring may be placed anywhere on the bolt thereby enabling one to obtain any desired spring pressure. Since the ring is installed by hand, it is difficult to exert sufficient pressure to cause the glass spacers to crack, a condition prevalent with the threaded bolt.

All parts are inexpensive and easy to use. The glass parts are manufactured in the local glass shop, and the spring and retaining ring are commercially available.

1. LC-022-2-SS, Lee Spring.
2. 5555-18, Waldes Truarc.

FIG. 1. Breakdown showing pyrex glass bolt, spring, and retaining ring. On the left is an assembly of plates held together by means of these bolts.

Three distinct advantages are readily apparent in the use of this new assembly. The components are assembled and disassembled with much greater ease in the hoods, parts breakage has been eliminated, and with a stronger spring and smaller diameter glass spacers (7-mm tubing) than supplied by the manufacturer, a more rigid and better aligned assembly has resulted.

3.3   Ultrapurification of Graphite by Preburning for Use in Emission Spectrography and Solids Mass Spectrography

R. K. Skogerboe, A. T. Kashuba, and

G. H. Morrison

Chemistry Department, Cornell University,
Ithaca, New York 14850

Preburning graphite spectrographic electrodes in an arc to reduce the residual elemental impurity levels is a common and effective practice. In emission spectrography, it is generally considered desirable to preburn the electrodes for a time at least equivalent to the anticipated exposure time to eliminate the electrode blank. This is a routine practice in the authors' laboratory for trace analysis problems. It has been observed that impurity levels can be reduced to the nondetectable level (by emission spectrography) by preburning in a 35-A dc arc under a rare gas atmosphere for 30–40 sec.

When nonconducting or micro samples are to be analyzed by spark source mass spectrography, graphite electrodes or powder are also used.[1-3] The ultimate sensitivity of this technique may be as much as three orders of magnitude better than that of emission spectrometry so the electrode blank problem is certainly more serious. Brown and Wolstenholme,[3] for example, report six elements present in graphite at the 3 ppm atomic (ppma) level, four at 1 ppma, and 16 at levels between 0.3 and 0.01 ppma by spark source mass spectrography. Considering the fact that these blank levels would complicate, if not preclude, the determination of a number of trace elements in samples mixed with or contained in graphite, one might conclude that another conducting powder would solve the problem. The fact remains, however, that graphite, in addition to being relatively inexpensive and easy to handle, is one of the most pure conducting materials available.[2,3] It is also one of the few materials that offers the possibility of convenient and simple further purification in the laboratory. Thus, a study of graphite purification by preburning was performed to

1. J. W. Guthrie, Analysis of Special Samples in *Mass Spectrometric Analysis of Solids*, A. J. Ahearn, Ed., (Elsevier Publishing Co., New York, 1966).

2. E. B. Owens, ASTM E-14, G. A. M. S. Brit. Inst. Pet. Meeting, Paris, (Sept. 1964).

3. R. Brown and W. A. Wolstenholme, ASTM E-14 Meeting on Mass Spectrometry, San Francisco (1963), Paper No. 75.

determine the conditions necessary to reduce residual impurities to the lowest possible levels.

In brief, preformed graphite electrodes (National AGKSP grade) were preburned for various time intervals in the enclosed chamber described by Rupp, Klecak, and Morrison.[4] A 35-A (shorted electrode) dc arc was used under a dynamic atmosphere of argon flushing at 6 L/min. The chamber design is such that it is easily mounted in the electrode holders of the arc stand and prevents the back diffusion of air into the arc column. The choice of the high current density discharge was dictated by emission spectrographic observations. By using an inert gas atmosphere, the electrodes can be extensively preburned with only small weight losses and no visual changes in shape. Graphite powder (National SP-1C) was preburned after packing approximately 50–70 mg into a previously preburned preform (National # L-3900).

After preburning, the preformed electrodes and the powder electrodes formed by pressing in a Seidel mould were presparked in the mass spectrograph to remove surface contaminants which might have been picked up during transfer to the system and subsequently analyzed by conventional procedures previously described.[5] The samples that were not preburned were handled similarly. The resulting photoplates were densitometered and exposure ratios (using ^{13}C as the reference) were computed by normal means.[4,5] Sensitivity corrections were made for the elements for which the manufacturer provided a batch analysis. Unit sensitivity was assumed for elements not detected by the manufacturer. Either approach is open to criticism so it may be advisable to consider the concentrations as order of magnitude estimates in terms

4. R. L. Rupp, G. L. Klecak, and G. H. Morrison, Anal. Chem., **32**, 931 (1960).
5. W. L. Harrington, R. K. Skogerboe, and G. H. Morrison, Anal. Chem. **37**, 1480 (1965).

Table I. Effect of preburning on impurity concentration (ppm by wt).[a]

| Element | Preburn condition (min) | | | | | |
	0	1.5	3.0	3.0 +LiCO$_3$ flux	5.0	9.0
Al	0.7	0.006	0.003	0.003	0.001	0.001
Ca	3	0.1	0.03	0.3	0.0015	0.003
Cl	3[b]	0.02	0.02	0.3	0.003	0.003
Cr	1.0[b]	0.08	0.08	0.02	0.08	0.05
Cu	0.1	0.009	c	0.01	0.007	0.006
Fe	2	0.25	0.1	0.2	0.1	0.1
Ni	0.3[b]	0.03	0.015	0.03	0.009	0.006
N	3[b]	0.05	0.006	4	0.003	c
O	8[b]	0.04	0.006	12	0.03	c
K	3[b]	0.003	0.001	0.003	<0.001	<0.001
Li	3[b]	0.04	0.007	1.5	0.009	0.006
Na	5[b]	0.002	0.0015	0.005	0.0015	<0.001
S	5[b]	0.015	0.007	0.08	0.015	0.007
V	0.3[b]	0.003	0.002	0.04	0.003	0.003
Zn	0.2[b]	0.02	0.01	0.01	0.01	0.01

[a] Mg and Ti not determined due to Ca$^+$ interference.
[b] Concentrations estimated assuming unit sensitivity. All other concentrations corrected for sensitivity using the batch analysis provided by the manufacturer.
[c] Quantitation precluded by photoplate defect or background halation.

of absolute accuracy. The relative concentrations may also reflect impurity differences in the individual electrode pairs used for each preburning time but the comparisons should be generally indicative of the degree of impurity reduction.

In the case of the preformed electrodes, the data in Table I indicate that impurity levels can be reduced by a factor of 1000 or more by preburning in excess of 5 min. In fact, the majority of the impurities detected can be reduced to the 1–10 ppb level which is essentially the limit of detectability for the technique. The data also implies that, for elements such as Cr, Fe, Ni, and V; the tendency to form refractory carbides may limit the degree of purification. The fourth and fifth columns of Table I indicate the effect of using a fluxing material (spectrographically pure Li_2CO_3) as an aid to purification. In this instance, 10 mg of Li_2CO_3 were totally vaporized from the electrode cup after approximately 1.5 min of preburning. It appears, however, that the presence of the flux reduces the degree of cleanup for several elements. If these elements are present in the flux, their levels are below the emission spectrographic detection limits so it is most reasonable to assign the temperature reduction caused by the flux vaporization as the cause.

Although purification is observed, the extent is less pronounced for the graphite powder, data given in Table II. The lack of reduction in the levels for Cr, Fe, and Ni can most probably be attributed to contamination of the powder by the stainless steel die used for the formation of the electrodes.

It is apparent that rigorous prolonged preburning is required to achieve maximum graphite purification.

Table II. Effect of preburning on graphite powder impurity levels (ppm by wt).[a]

	Preburn time (min)	
Element	0	5
Al	0.1	0.02
Ca	2[b]	0.2
Cl	3[b]	0.05
Cr	0.4[b]	0.5
Cu	0.1[b]	0.01
Fe	0.1	0.3
Ni	0.05[b]	0.05
N	25[b]	0.01
O	40[b]	0.02
K	4[b]	0.08
Li	0.3[b]	0.04
Na	5[b]	0.06
S	15[b]	2
V	0.1[b]	0.08
Zn	0.05[b]	0.01

[a] C_n^+ interference precluded determination of Mg and Ti.
[b] Concentrations estimated assuming unit sensitivity.

Although elemental levels can be reduced to below emission spectrographic detection limits by shorter preburn times, this does not preclude the possibility of blank contributions particularly for analyses at low trace concentration levels. Hence, longer preburn times are suggested for both emission and mass spectrographic usage.

Research supported by the Advanced Research Projects Agency.

SECTION 4
NUCLEAR MAGNETIC RESONANCE

Microcell for Nuclear Magnetic Resonance Analyses

4.1

R. A. Flath, N. Henderson, R. E. Lundin, and
R. Teranishi

*Western Regional Research Laboratory, Agricultural Research
Service, U. S. Department of Agriculture, Albany, California
94710*

The use of a commercially available all-glass spherical sample-chamber microcell in combination with a time-averaging computer for nuclear magnetic resonance (NMR) analyses of microsamples has recently been reported.[1] Brame[2] has described a technique for trapping small gas-chromatographically isolated samples directly in such an all-glass NMR microcell. A somewhat different, easily fabricated microcell has been constructed and tested in our laboratories with excellent results. The design is a modification of a cell first described by Frei and Niklaus.[3]

1. R. E. Lundin, R. H. Elsken, R. A. Flath, N. Henderson,
   T. R. Mon, and R. Teranishi, Anal. Chem. **38**, 291 (1966).
2. E. G. Brame, Jr., Anal. Chem. **37**, 1183 (1965).
3. K. Frei and P. Niklaus, private communication.

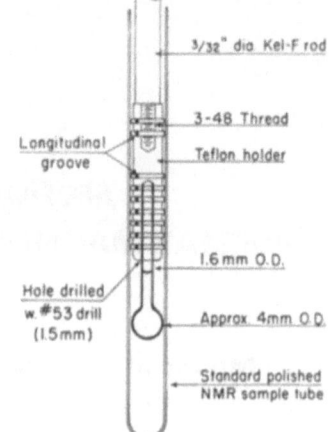

FIG. 1. Microcell for NMR.

The microcell, shown in Fig. 1, consists of a standard NMR sample tube, a machined Teflon plug,[4] and a small glass bulb in which the sample is placed. The spherical sample bulbs are prepared from standard borosilicate melting-point tubing. Usually the diameter of the bulb is formed so as to provide a smooth slip fit in the standard NMR tube; this yields a microcell holding approximately 35 μl of solution. The Teflon holder is turned on a small lathe to provide a snug fit in the standard NMR sample tube. (This snug fit and the considerable difference in coefficients of thermal expansion of Teflon and borosilicate glass will preclude the use of the microcell at elevated temperatures.) A 7-mm-deep hole is drilled into the lower end of the plug to accept the neck of the microsample bulb. This hole must be of sufficiently small diameter to grip the neck of the sample bulb firmly and must be concentric with the

4. Reference to a company or product name does not imply approval or recommendation of the product by the U. S. Department of Agriculture to the exclusion of others that may be suitable.

body of the Teflon holder in order to minimize sample wobble while spinning. A longitudinal groove is cut into the plug to equalize pressure above and below the plug when it is placed in the NMR tube. The sample bulb and Teflon plug are positioned in the NMR tube for maximum resolution by means of the threaded Kel–F rod, which screws into a threaded hole in the top of the Teflon plug. The rod is removed before the cell is placed into the instrument probe. The space around the microsample bulb, below the Teflon holder, is filled with carbon tetrachloride to minimize further any tendency for the sample bulb to wobble when the assembly is rotated.

This simply constructed microcell has several advantages over commercially available microcells. The sample bulbs have short necks (approximately 15 mm), which make filling and cleaning very convenient, without the use of special, long-needle, large dead-volume syringes. The bulbs are quite inexpensive and can be prepared easily in any volume up to 35 μl from melting-point tubing selected to fit snugly in the Teflon holder (a hollow-shaft, variable-speed stirring motor makes a convenient glass lathe for blowing spherical bulbs). The most attractive feature of the microcell system described is the increase in resolution obtainable, particularly with the 100-MHz instrument. This higher resolution is presumably obtainable because the heavy variable-thickness walls of the commercial capillary microcell are replaced by the thin walls of a standard NMR tube and the very thin, even walls of the microbulb. An example of a single scan analysis on the 100-MHz instrument with 0.45 mg (0.50 μl) of *trans,trans*-humulene[5,6] in 34 μl of carbon tetrachloride is shown in Fig. 2. Usable

5. A. T. McPhail, R. Reed, and G. A. Sim, Chem. Ind. (London) **1964**, 976 (1964); J. A. Hartsuck and I. C. Paul, *ibid.*, p. 977.
6. Sample kindly provided by R. G. Buttery.

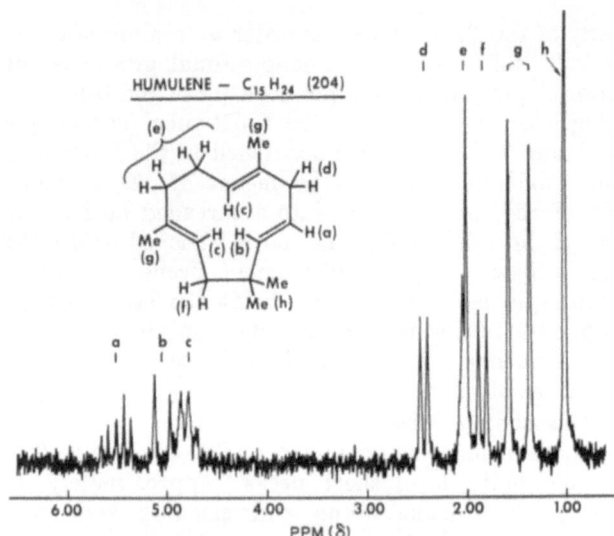

FIG. 2. Single scan spectrum of *trans,trans*-humulene; 100 MHz, 0.45 mg, in 34 μl of carbon tetrachloride.

single scan spectra with 1–2 μM of hydrogen can be obtained, even when the absorption appears as a multiplet. The high resolution possible with this type of microcell permits longer time-averaging runs without serious deterioration of the resultant spectrum resolution.

The simple construction, easy sample-chamber accessibility, small volume, and high resolution make this microcell useful to any investigator attempting NMR analyses of microsamples.

Microcells for Nuclear-Magnetic- 4.2
Resonance Spectroscopy

S. C. Slaymaker

Shell Oil Company, Houston Research Laboratory,
P. O. Box 100, Deer Park, Texas 77035

One major preoccupation of nuclear-magnetic-resonance (NMR) spectroscopists has been to minimize the size of samples required in order to obtain usable nuclear-magnetic-resonance spectra.

The basic limitation to the sensitivity of nuclear-magnetic-resonance lies in the signal-to-noise ratio of the equipment used. When limited quantity of sample is available and time averaging equipment is lacking, the available sample must be utilized in the most efficient manner. For this reason, a number of microcells have been developed privately and commercially for NMR work. In general, the spherical sample cell appears to be the most popular because it approximates closely to the theoretical Lorentzian cavity. Deviations from an ellipsoidal shape lead to a loss in resolution due to distortion of the magnetic field. Differences in the bulk magnetic susceptibility of the material from which the cell is constructed and the solvent and/or sample accentuate this distortion effect. The various theoretical aspects of this subject are discussed more fully in the literature.[1-3]

A second cell, which approaches the requirements of the ellipsoidal shape, is a cylinder having an infinite ratio of length to diameter. In practice, it has

1. A. A. Bothner-By and R. E. Glick, J. Chem. Phys. **26**, 1647 (1957).
2. W. C. Dickinson, Phys. Rev. **81**, 717 (1951).
3. J. R. Zimmerman and M. R. Foster, J. Chem. **61**, 282 (1957).

been found that length of approximately five times the diameter of the receiver coil is sufficient for most purposes,[3] although this is possibly due to the fact that the ends of the sample column are out of the effective range of the receiver coil rather than to a good approximation of an ellipsoidal shape. This type of cell has become the accepted standard for NMR work, but it has the disadvantage that a considerable portion of the sample is situated in the ends of the tube remote from the receiver coil, where it is ineffective for signal production.

We have found that very fine bore capillaries serve as effective microcells for liquid samples or for concentrated solutions in which the entire content of the capillary contributes to the NMR signal. Such tubes, having inside diameters of about 0.025 to 0.12 cm and containing from 0.5 to 10 μl/cm, are inexpensive and give added dividends in resolution, convenience, and time.

The tubes are filled by capillary action, sealed at one or both ends by a small flame, and are dropped into regular 5-mm o.d., thin-walled, NMR spinning sample cells. When using the Varian A-60 NMR spectrometer, the optimum length of sample column was found to be approximately one centimeter. Shorter sample lengths resulted in weak signals in the smaller capillaries and loss of resolution in the large ones. On the other hand, longer sample lengths did not result in proportionately larger signals.

The sample cells containing the capillaries are adjusted by raising or lowering them in the probe for maximum signal. On spinning, the capillary was observed to ride up the rounded bottom of the thin-walled cell and to take up a position against the flat inner wall of the outer tube. This made the adjustment of sample position somewhat dependent upon spinning rate. In theory, since the sample revolves in a circular orbit around the axis of spin, a heavy-walled outer tube should have improved resolution, since the sample then frequented a smaller volume inside the coil. In practice, we have noted no improvement in resolution upon using the heavy walled NMR Specialties semimicrocell as a container for the capillary. This is probably due to the fact that the observed resolution is limited to the line-width allowed by the high filtering level required to control noise.

Table I lists various sample cells investigated together with a measure of resolution and comparative efficiencies measured under identical machine parameters. As will be seen, the spherical cells have the highest efficiency rating; however, a considerable loss of resolution has been observed in most cases. The capillary cells showed a three-fold efficiency improve-

Table I. Efficiency and resolution of various sample cells.

	Ethyl-benzene content (μl)	S/N[a]	E[b]	Line width[c]
Varian sensitivity standard	4.0	7.6	1	1.9
Spherical plastic microcell	1.0	8.1	4.2	3.0
Spherical glass microcell	1.0	6.6	3.4	2.4
0.027-cm diam capillary tube	0.5	3.1	3.3	1.8
0.045-cm diam capillary tube	1.5	8.4	2.9	1.7

[a] The signal-to-noise ratio is arbitrarily defined as the height of the largest peak in the ethylbenzene quartet divided by two-fifths of the maximum pen excursion over a span of two inches prior to the ethylbenzene quartet.
[b] The cell efficiency E is defined as four divided by the number of microliters of ethylbenzene required to give a signal-to-noise ratio equal to that of the Varian Sensitivity Standard.
[c] The line width is defined as the half-height line width of the center peak of the ethylbenzene triplet. The line width of 1.8 cps is determined by the high degree of filtering rather than by field homogeneity.

ment over the standard Varian sample tubes and
showed no resolution loss. The greatest advantage of
the capillary cells resulted from the fact that, at least
with liquid samples, dissolution is unnecessary, and, if
the sample is received in a sealed capillary, a spectrum
can be obtained directly without any form of sample
manipulation. This feature makes the procedure particularly valuable and convenient for the examination of gas chromatographic cuts, etc.

Recent experiments by J. M. Martin, Jr. of this
laboratory have demonstrated the compatability of
these capillary cells with time-averaging techniques.
Efficiencies greater than 40 have been observed using
0.027 cm capillary tubes in conjunction with a Mne-
motron computer of average transients.

4.3   Microsampling Techniques for Combined Gas Chromatography and High-Resolution Nuclear Magnetic Resonance Spectroscopy

B. Milazzo, L. Petrakis, and P. M. Brown

*Gulf Research & Development Company,
Pittsburgh, Pennsylvania 15230*

The realization of the great potential of high
resolution NMR in the elucidation of molecular structure has been accompanied by considerable progress
towards overcoming its relative lack of sensitivity.
The need to record the spectra of the individual
components of mixtures and of samples available in
very small quantities has led to the utilization of the
gas chromatograph in conjunction with NMR, to
development of microcells for the optimum use of

available samples, and to the use of computers to improve signal-to-noise ratio.[1-5]

We wish to report on an arrangement that we have used successfully to record the spectra of micro-samples and of individual components of mixtures. Our method equals or exceeds the resolution and sensitivity of other reported methods, and it is simple, quick, and requires no extensive preparation or expensive equipment. It eliminates the need for costly, cumbersome microcells,[3,4] and overcomes the drawbacks in time and sample required of the trap and transfer method[1,2] and the adjustment difficulties of the free-floating capillaries.[5]

The simple design of the easily constructed sample holder is shown in Fig. 1. It consists of a standard thin-wall NMR tube at the bottom of which a 2-mm hole has been bored, and a capillary tube held inside the standard tube with the aid of precision-machined Teflon spacers. The capillary tube was chosen because of its lower cost, greater resolution and sensitivity, structural uniformity, handling ease, and ready adaptability of chromatographic trapping. A 1.7-mm o.d., 1.3-mm i.d. capillary was cut to a 6-cm length. This short length proved practical and overcame condensation difficulties encountered in using the much longer traps and microcells. Two 2.5-cm length Teflon spacers were precision machined for a loose-sliding fit in the standard NMR tube. A 1.8-mm diameter hole bored through the center of these spacers held the capillary in a tight sliding fit. A

1. D. E. Archer, J. H. Shively, and S. A. Francis, Anal. Chem. **35**, 1369 (1963).
2. J. G. Bendoraitis, B. L. Brown, and L. S. Hepner, Anal. Chem. **34**, 49 (1962).
3. E. G. Brame, Jr., Anal. Chem. **37**, 1183 (1965).
4. R. A. Flath, N. Henderson, R. E. Lundin, and R. Teraniski, Appl. Spectry. **21**, 183 (1967).
5. C. S. Slaymaker. Appl. Spectry. **21**, 42 (1967).

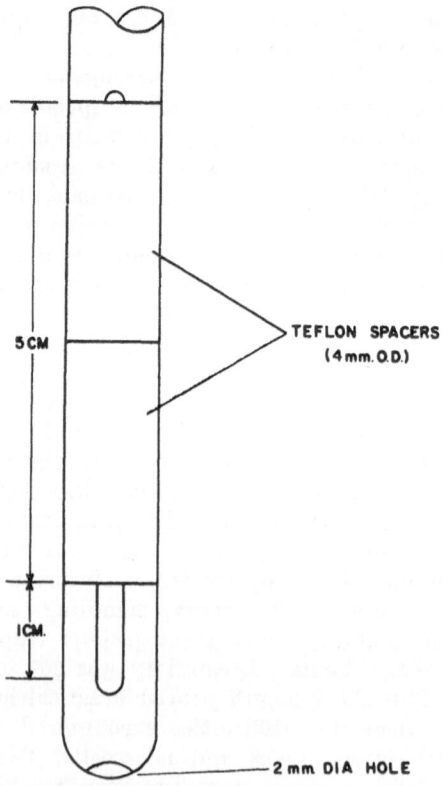

FIG. 1. NMR microcell showing the positioning of capillary in standard NMR tube.

2-mm hole bored through the bottom of the standard tube eliminated cumbersome adaptors and fulfilled two requirements: (1) equalization of pressure above and below the Teflon spacers in the standard tube; and (2) a simple, easy method of adjusting and removing the sample capillary by pushing a wire rod through the hole. The optimum length of sample column was found to be between 0.5 and 1 cm.

Shorter sample lengths result in weak signals and a loss of resolution. Therefore, dilution is necessary for sample volumes less than 5 μliters (μl). Also, before placing the entire assembly into the NMR probe, the capillary is adjusted so that the receiver coil will be directly in the center of the sample column.

The chromatographic trapping was accomplished by adapting a hypodermic needle to the exit port of a gas chromatograph. The hub of the hypodermic needle (Hamilton N-73529) was carefully removed using a vise and pliers. The blunt end was inserted through a solid septum and attached to the exit port using a standard septum nut. This type of adaptation has been successful for Varian-Aerograph and F&M chromatographs. Just before the desired peak is eluted, the needle is heated using a heat gun to remove traces of any previously eluted peak and also any column stationary phase. Our experience to date indicates that compounds as low boiling as 133°C and separated by only 9°C in bp may be collected with the 6-cm capillary. Longer capillaries would permit trapping of lower boiling compounds. Compounds which are solid at room temperature or which freeze in the cooling bath may be expected to present some difficulty.

All spectra were recorded on a Varian HA-60 spectrometer operating in the external lock mode. Good spectra have been obtained routinely with 1–5 μl samples. Figure 2 shows the spectrum of the 1 μl of ethanol diluted to a total volume of 5 μl in chloroform. The resolution is 2.3 Hz and the S/N equals 57. The rigid positioning of the capillary resulted in increased resolution and S/N ratio, without the extensive and laborious adjustments required for the free-floating capillary. Time averaging allows the

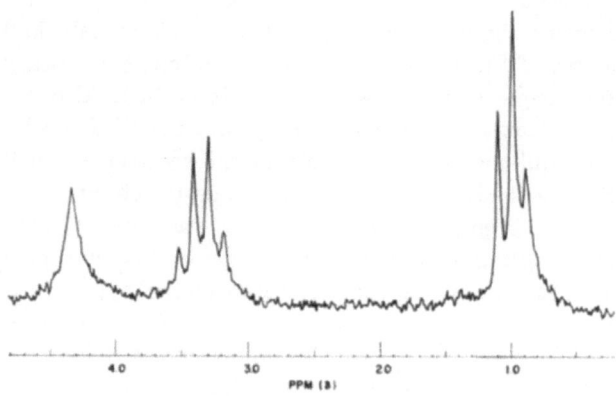

FIG. 2. Spectrum of 1 μl of ethanol obtained with microcell and no time averaging.

recording of spectra of even smaller quantities. For example, the spectra of 0.1 μl of n-propyl benzene have been recorded after time averaging 100 scans over a 3-h period, demonstrating the excellent stability of this simply designed sample holder. In the course of operations, quite acceptable integrals have been taken on the microsamples using the Varian C-1024 time-averaging computer; also, spin-decoupling experiments have been carried out, and the whole sample assembly has been heated up to 80°C in the course of identifying OH bands. During all these operations the microsample holder has yielded very satisfactory results.

Thus, this fast, practical method utilizing the simple sample holder and ordinary capillary tubes allows the examination of microsamples or of individual components of mixtures simply and efficiently and further extends the great analytical potential of high resolution NMR. For spectrometers operating in the internal lock mode and/or at 100

MHz the inherent greater sensitivity and longer sweep times possible should allow one to record routinely the NMR spectra of one or less microliter samples without time averaging.

"Multi-Cloistered" NMR Cells   4.4

E. M. Banas

Research and Development, American Oil Company,
Whiting, Indiana

Special "multi-cloistered" NMR cells have been developed to simplify analyses of stereoisomers and other compounds whose NMR spectra exhibit very slight differences in chemical shifts. They are also applied in studies involving solvent effects.

As shown in Fig. 1, the cells comprise banks of small tubes held immobily within larger ones. Ordinarily they are made of standard stock items. For example, our quadra-cloistered cell, Fig. 1(a), comprises

FIG. 1. (a) quadra-cloistered cell; (b) hepta-cloistered cell.

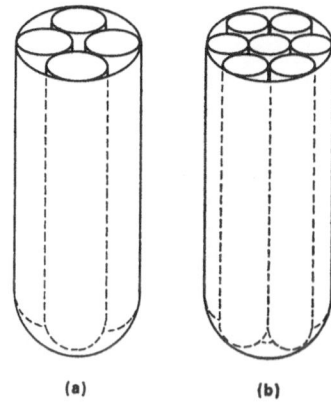

(a)   (b)

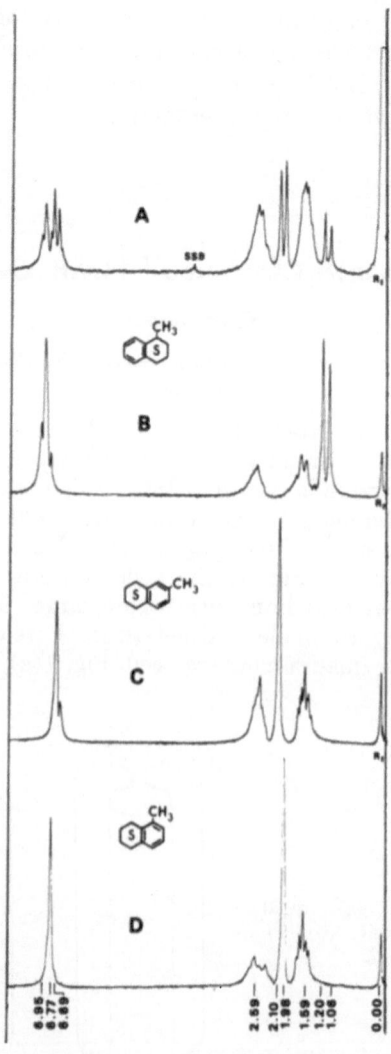

FIG. 2. A, composite spectrum of methyltetralin isomers B, C, and D, and TMS, obtained with a quadra-cloistered NMR cell.

a standard 5-mm NMR tube containing four Kimax No. 34505 melting point capillaries that have been sealed at one end. Similarly, the heptacloistered cell, Fig. 1(b), is made of a 5-mm tube containing seven Fisher-Scientific No. 5-962 coagulation capillaries. Other combinations are possible.

For the analyses of two or more samples, a cell containing the appropriate number of tubes is filled, placed in a homogeneous field, and spun. Relative chemical shifts are thus recorded together on the same graph. With the quadra-cloistered cell, for example, we find that we can ignore effective external magnetic field effects, susceptibility corrections, intermolecular association effects and shapes of the sample. As a result, large numbers of samples can be analyzed very rapidly.

Figure 2 shows a typical analysis, where A, the composite spectrum of three isomers of methyltetralin, is compared with the individual spectra (at 60 MHz and referenced externally). On occasion, we have been able to detect shift differences of 1 part in 10^8.

Apparatus for the Preparation of Sealed Samples for NMR

4.5

William J. Lanum and Francis R. McDonald

U. S. Department of the Interior, Bureau of Mines,
Laramie Petroleum Research Center, Laramie, Wyoming 82070

Many times when measuring the NMR spectra of certain compounds it is necessary to protect the sample from atmospheric oxygen or water. The effect of dissolved oxygen on the NMR spectrum is well known.[1] It may also be necessary or desirable to

1. See, for example, J. W. Emsley, J. Feeney, and L. H. Sutcliffe, *High Resolution Nuclear Magnetic Resonance Spectroscopy* (Pergamon Press, Ltd., London, 1965), Vol. I, p. 260.

obtain the NMR spectrum with the sample compound in an appropriate solvent together with a reference standard. These considerations are particularly important when obtaining reference spectra of high-purity standard-sample compounds.

This laboratory is engaged in obtaining reference NMR spectra of API–USBM Standard sulfur compounds and nitrogen compounds. To prepare these samples properly, it is necessary to seal a measured volume of each compound together with a measured amount of solvent and reference compound in an NMR sample tube while excluding air. It is also convenient to prepare and store a number of samples in advance, such that the standard compound is isolated from the solvent and reference compound until just prior to obtaining the spectrum.

The literature did not reveal any convenient method for meeting our sampling requirements. The simple all-glass apparatus described in this paper fulfills all of the requirements and yet is easy to construct and use.

Description and Use of Apparatus: Figure 1 is a diagram of the apparatus. The all-glass unit consists of two chambers (A and B) separated by a break-off seal. A solid glass bar, for breaking the seal, is enclosed in chamber B and the NMR sample tube is sealed onto a side arm on chamber B.

Chamber A, the sample chamber, is sealed onto a vacuum manifold by means of the side arm. A measured amount of sample is vapor transferred into the chamber and sealed off. The apparatus is then carefully inverted (care must be exercised to avoid damaging the break-off seal) and resealed onto the vacuum manifold by means of the remaining side arm on Chamber B. Measured quantities of solvent and NMR reference standard are then vapor transferred into chamber B and the apparatus is again sealed off.

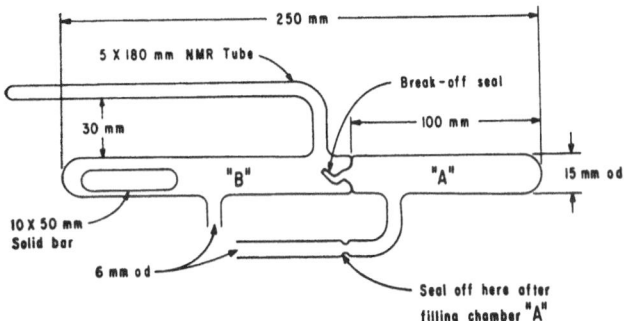

FIG. 1. Apparatus for preparation of sealed NMR tubes.

The sample, solvent, and reference standard are isolated from each other as well as from the atmosphere and may be stored for future use.

To prepare the sample for use in the spectrometer the break-off seal separating chambers A and B is broken, using the solid glass bar in chamber B. Mixing the sample, solvent, and reference standard is best accomplished by cooling chamber A in liquid nitrogen and drawing the solvent and reference into chamber A. The solution may then be thawed and poured into the NMR tube which is then sealed off after being cooled in liquid nitrogen.

This apparatus provides the spectroscopist with a method of preparing a sample that has been isolated from the atmosphere, thus reducing the deleterious effect that may be brought about by exposure to oxygen, moisture, or other atmospheric agents. Samples subject to actinic decomposition could be protected by use of the proper type glass ampoule or storage in the dark.

The work upon which this report is based was done under a cooperative agreement between the Bureau of Mines, U. S. Department of the Interior, and the University of Wyoming.

SECTION 5
RAMAN SPECTROSCOPY

Polarized Raman Scattering from Amorphous Solids in Liquid Media

5.1

D. F. Shriver

*Department of Chemistry and Materials Research Center,
Northwestern University, Evanston, Illinois 60201*

Raman polarization data, which are obtainable on gas, liquid, and oriented single crystal samples, provide a powerful tool in the study of molecular vibrations and molecular symmetry. However, this type of data has been unobtainable for finely divided solids because reflection and refraction at the solid surfaces leads to depolarization of the incident and Raman scattered light. In principle, it should be possible to eliminate this depolarization by immersing the solid in a medium of identical index of refraction. To our knowledge this technique has never been applied to obtain polarization data but it has been employed to increase the efficiency of Raman scattering.[1-3] Also, immersion of finely divided solids in a medium of similar index of refraction has been used to reduce light scattering in absoption spectros-

1. J. L. Cojan and A. Renaux, Compt. Rend. **247**, 1111 (1958).
2. M. Tobin, J. Opt. Soc. Am. **49**, 850 (1959).
3. R. Savoie and J. Tremblay, J. Opt. Soc. Am. **57**, 329 (1967).

copy experiments.[4] In this note, attention will be confined to Raman scattering from amorphous solids, silica gel, and an ion exchange resin.

A slightly basic, nearly saturated solution of Na_2CrO_4 was added to previously dried (100°) finely divided (<100 mesh) wide pore silica gel (Davison No. 70) to produce an apparently dry solid containing 45% by weight of the solution. Raman spectra were obtained on a double monochromator instrument (SPEX) in conjunction with 6328.17 Å He–Ne laser excitation and photon counting detection.[5] The 90° illumination geometry was employed with the focused laser beam impinging on the front surface of the sample (in a fused quartz or glass container) at a grazing angle. The chromate ion was chosen for this initial work because it is an excellent Raman scatterer. Spectra in the ν_1, ν_3 region of CrO_4^{2-} are presented in Fig. 1. It is clear from this figure that in the absence of a liquid medium the polarization is negligible; however, immersion of the solid in hexane leads to a low depolarization ratio for ν_1 ($\rho_s < 0.1$) and approx. $\rho_s = 0.75$ for ν_3. In addition ν_2 and ν_4 are observed at 355 and 375 cm^{-1}. These bands overlap and upon rotation of the polaroid analyzer they coalesce into one broad band. The depolarization ratio for the combined band envelopes is 0.8 which is close to the value expected for randomly oriented CrO_4^{2-} groups (0.75).

Another example is afforded by a sample of air-dried Dowex 1 8X (100–200 mesh) in the CrO_4^{2-} form which was investigated in the region of ν_1. When the solid was immersed in a medium of similar index of refraction (benzene) spectra were obtained in which ν_1 appeared to be totally polarized. This peak is sufficiently separated from a polarized peak

4. R. L. Burwell, Jr., R. G. Pearson, G. L. Haller, P. B. Tjok, and S. P. Chock, Inorg. Chem. 4, 1123 (1965)
5. I. Wharf and D. F. Shriver, Inorg. Chem. 8, 914 (1969).

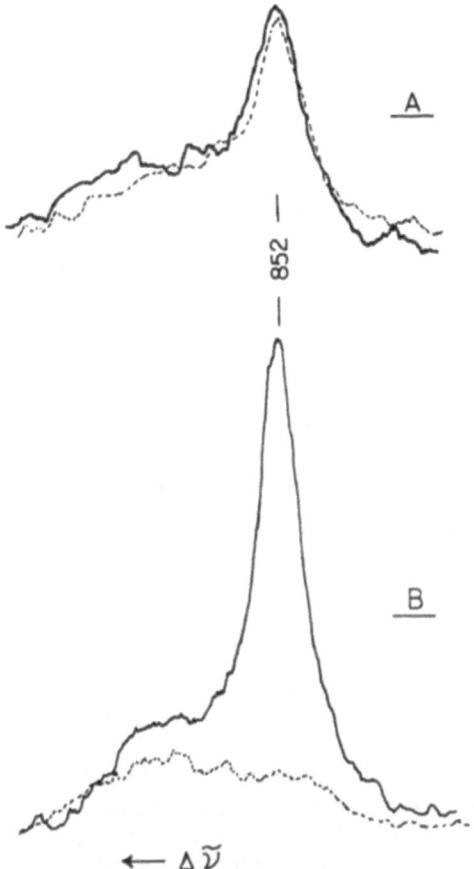

852

A

B

← $\Delta \tilde{\nu}$

FIG. 1. Raman Spectra of CrO_4^{2-} in Silica Gel. (ν_3 appears as a high $\Delta \bar{\nu}$ shoulder on ν_1) (A) "Dry" solid; $I_{||}$ (obs. $\perp$)—; $I_\perp$ (obs. $\perp$) – – – (B) Solid immersed in hexane; $I_{||}$ (obs. $\perp$) —; $I_\perp$ (obs. $\perp$) – – –. The polarization notation is that of E. B. Wilson, Jr., J. C. Decius, and P. C. Cross, *Molecular Vibrations* (McGraw–Hill Book Co., New York, 1955), p. 45. The position of ν_1 in cm^{-1} is indicated in the figure.

of the resin (approx. 787 cm^{-1} in the Cl$^-$ form of
the resin) to avoid confusion.

The depolarization ratio of a species of intermediate symmetry, $S_2O_3^{2-}$ was studied using Ar-ion laser
excitation ($\lambda = 5145$ Å). In solution, the depolarization ratio of the lowest totally symmetric mode, ν_3,
was found to be 0.16. When this saturated solution
was adsorbed on silica gel and the resulting solid immersed in heptane (as described above), a depolarization ratio of 0.14 ± 0.06 was observed. This reasonable
agreement demonstrates the utility of the present
technique for low symmetry species, which have non-
zero depolarization ratios for symmetric modes.

It is concluded that polarization data may be obtained on finely divided amorphous substances and
that the depolarization ratios will approximate those
expected for randomly oriented molecules. High precision is not to be expected from the present technique because it is virtually impossible to eliminate
all sources of extraneous scattering such as internal
cracks in solid particles and minute air bubbles.
However, the results are sufficiently precise to form
the basis of assignments and characterization. Therefore, the technique should be quite useful for the
characterization of species in adsorbants, ion exchange resins, and similar amorphous solids. When
finely pulverized crystals are observed using the
present technique, polarization data are obtainable
but the rules governing the polarization are different
from those based randomly oriented molecules. The
details are currently under investigation.

Finally, it is important to note that the index of
refraction match between the solid and liquid is not
critical. In the experiments described above, the
match was achieved by visual observation of the solid
in several organic liquids. The liquid which gave
the clearest slurry was picked for the study. To investigate the influence of refractive index mismatch,

the $S_2O_3{}^{2-}$–H_2O-silica gel system was immersed in 50% benzene–heptane, with the result that the slurry had a much more milky appearance than observed with pure heptane. The depolarization ratio for ν_3 of $S_2O_3{}^{2-}$ under these poorer conditions was 0.2. Since the difference in the refractive index between heptane and the benzene–heptane mixture is sizable (0.043), this result clearly indicates the latitude available for picking the liquid medium.

This research was supported by the Advanced Research Projects Agency through the Northwestern University Materials Research Center, and by an Alfred P. Sloan Fellowship to the author. I am grateful to R. C. Taylor for discussions on Raman scattering of solids and to R. L. Burwell, Jr., for discussions concerning adsorbed species.

Variable Temperature Raman Cell   5.2

D. J. Antion* and J. R. Durig

*Department of Chemistry, University of South Carolina,
Columbia, South Carolina 29208*

A number of cryostats[1-5] have been designed previously for use with the Cary 81 Raman spectrophotometer equipped with the Toronto arc source. The advent of laser Raman spectroscopy has neces-

*National Science Foundation predoctoral fellow.
1. J. R. Ferraro, J. S. Ziomek, and K. Puckett, Rev. Sci. Instr. **35**, 754 (1964).
2. J. E. Griffiths, R. P. Carter and R. R. Holmes, J. Chem. Phys. **41**, 863 (1964).
3. N. Craig and J. Overend, Spectrochim. Acta **20**, 1561 (1964).
4. R. Savoie and A. Anderson, J. Opt. Soc. Am. **55**, 133 (1965).
5. J. I. Bryant, Spectrochim. Acta **24A**, 9 (1968).

sitated the development of new sampling techniques and new cryogenic devices. Two cryostats which permit 90° viewing of the Raman scattering were developed for laser excitation by Gee and O'Shea[6] and by Perry.[7] Carlson[8] has described a fixed-temperature Raman cell which is suitable for studying solids and condensed vapors with the laser source. We wish to describe a variable temperature cell which is also used in conjunction with the laser source. The versatility of this new Raman cell is exemplified by the fact that Raman spectra of liquids, solids, and condensed gases may be investigated over a wide temperature range.

The variable temperature cell [Fig. 1(A)] consists of three concentric Pyrex tubes. To the one end of the innermost tube (4-mm-o.d. sample tube) a 2-mm Pyrex light pipe is attached. The light pipe provides optical contact between the sample and the optical system of the spectrophotometer. The other end of the sample tube is sealed by a right-angle stopcock which permits deposition of vapors by means of standard vacuum line techniques. With the stopcock removed liquids and solids may be placed directly into the sampling area. The middle tube of the cell encloses the compartment through which cooled or heated nitrogen gas is passed. The outer tube comprises the vacuum jacket.

A special cradle [Fig. 1(B)] was designed and constructed for the variable temperature cell. The cradle is made of brass and possesses both horzontal and vertical adjustments for optimizing the sample position. Two adjusting screws which support the base of the cradle were machined to fit the V-slot inside the laser optics box. This arrangement per-

6. A. R. Gee and D. C. O'Shea, Rev. Sci. Instr. 37, 670 (1966).
7. C. Perry (private communication).
8. G. L. Carlson, Spectrochim. Acta 24A, 1519 (1968).

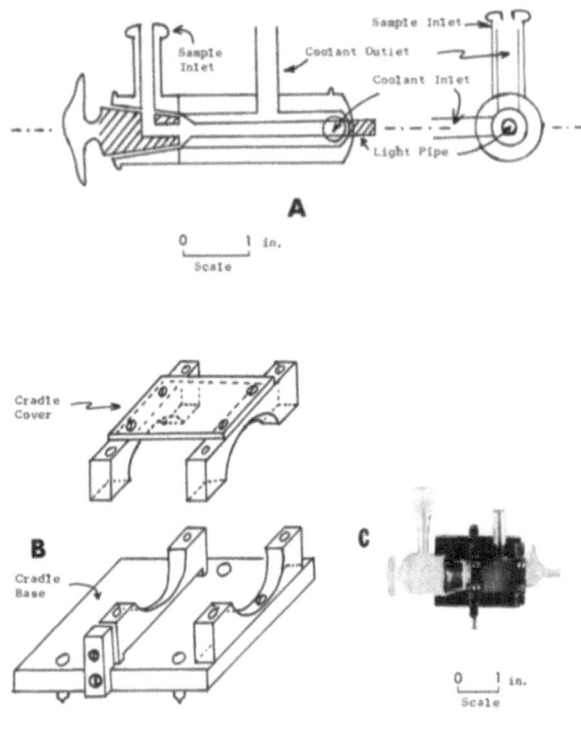

FIG. 1. Variable temperature Raman cell. (A) Cell schematic,
(B) cradle schematic, and (C) cell and cradle assembled.

mits removal and replacement of the entire assembly
[Fig. 1(C)] without loss of optimum conditions. The
over-all length of cell and cradle is approximately
12 cm and can easily be accommodated by the laser
optics compartment.

During cryogenic applications of the sample cell,
dry nitrogen gas is passed through a reservoir of
liquid nitrogen (b.p. −196°C) and from there into
the coolant jacket of the cell. The sample temperature

is variable over the range 23° to −178°C and is controlled by regulating the rate of flow of the coolant gas. Temperatures are measured with an iron–constantan thermocouple positioned in the coolant jacket through the exit port. Temperature gradients between a sample of carbon tetrachloride and the coolant were measured over the temperature range above and at no temperature did the difference exceed 10°C. Since the thermal conductivity of carbon tetrachloride is low (0.252×10^{-3} cal/cm sec deg) compared to that of Pyrex (1.63×10^{-3} cal/cm sec deg), the temperature difference of 10°C is probably a maximum. The sample and coolant thermocouples responded to a change in the flow rate within a few seconds and thermal equilibrium was attained at all temperatures in less than 5 min after the flow rate was altered. Constant flow rates (very small pressure variations) produced temperatures which were constant to ±2°C during a relatively long time interval (>1 h). Operating the sample cell above room temperature requires a heated reservoir, but the same simple procedure is followed.

A number of experiments were conducted in order to determine the optimum diameter of the light pipe. Light pipes of 6, 4, and 2 mm were affixed to sample tubes of 4-mm diam and the intensities of the Raman lines from liquid samples at room temperature were compared. For all cases studied the performance of the 2-mm light pipe exceeded the performance of each of the larger light pipes by as much as a factor of *five*. Light pipes whose diameters are less than 2 mm were not studied because of their fragile nature. The performance of the cell was tested with both liquid and solid samples. The Raman spectrum (below 500 cm^{-1}) of CCl$_4$ in a pyrex capillary tube was compared with the spectrum of CCl$_4$ in the sample cell at 23°C (Fig. 2). Under the same instrumental conditions the ratio of the in-

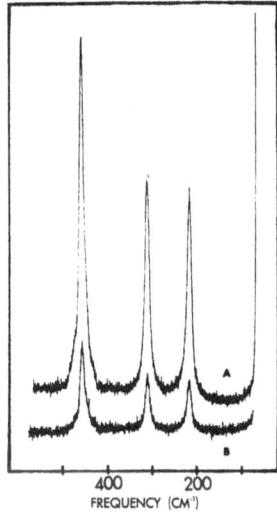

FREQUENCY (CM⁻¹)

Fig. 2. Raman spectrum of CCl₄ (below 500 cm⁻¹) taken under same instrumental conditions. (A) 1.6-mm Pyrex capillary tube, (B) variable temperature Raman cell.

tensities of the corresponding bands in the two spectra was never worse than 4 to 1 in favor of the capillary tube. With some light pipes the comparisons were comparable and we presume that the losses were dependent on the manner in which the light pipe was attached to the cell. The Raman spectra of some polycrystalline materials packed into the Cary solid sample holder were also compared with the spectra of the same compounds placed in the sample cell at 23°C. The reduction of the intensities of corresponding vibrational bands was a factor of two less than that found with the Cary solid sample holder.

To illustrate the variable temperature feature of the sample cell, the Raman spectrum of polycrystalline ammonium iodide was recorded over the temperature range 23° to −178°C. No low-frequency Raman data have previously been reported for ammonium iodide; however, neutron inelastic scattering

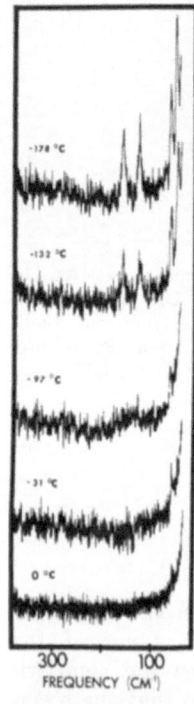

Fig. 3. Raman spectrum of NH₄I at various temperatures.

experiments[9-11] have shown that the librational fundamentals of the ammonium ion are centered near 294 cm⁻¹ at −173°C but the translational modes were not observed because the neutron data were only good down to 200 cm⁻¹. As the temperature of the sample of NH_4I was increased, the librational band broad-

9. G. Venkataraman, K. Usha Deniz, P. K. Iyengar, A. P. Roy, and P. R. Vijayaraghavan, J. Phys. Chem. Solids **27**, 1103 (1966).

10. A. Bajorek, K. Parlinski, and T. Machechina, *Inelastic Scattering of Neutrons* (International Atomic Energy Agency, Vienna, 1965), Vol. II, p. 335.

11. K. Mikke and A. Kroh, *Inelastic Scattering of Neutrons in Solids and Liquids* (International Atomic Energy Agency, Vienna, 1963), Vol. II, p. 237.

ened and disappeared at 300°K. There are no low-frequency fundamental vibrations in the Raman spectrum of NH_4I at room temperature but as the temperature was decreased, two bands appeared at 121 and 154 cm⁻¹. These bands became narrower and shifted to higher frequencies at lower temperatures. Figure 3 shows the Raman spectrum of NH_4I taken at various temperatures. These two Raman lines are probably translational lattice modes but the discussion of these bands in detail is the subject of a subsequent publication.[12] The variable temperature cell has also been used to investigate the Raman spectrum of oxalyl bromide, $[BrCO]_2$,[13] in which conformational isomers are thought to exist. In the Raman spectrum of the liquid there are too many vibrational bands to be accounted for on the basis of just one isomer. If the liquid is cooled below the melting point (−19°C), the more stable conformation of oxalyl bromide is expected to predominate and those bands in the Raman spectrum attributable to the less stable isomer should disappear as the equilibrium is shifted toward the more stable isomer. Figure 4 shows a portion of the spectrum of oxalyl bromide in which the Raman band at 407 cm⁻¹ exhibits considerable intensity at room temperature. As the temperature is decreased, this band decreases in intensity and eventually disappears. The conclusions regarding the existence of conformational isomers in oxalyl bromide are obvious and will be discussed fully in a future publication.[13] Further applications of the variable temperature cell are in progress in this laboratory.

We have presented a sampling device which can be used to observe 180° Raman scattering in condensed phases over a wide temperature range. The new cell is easy to fabricate and to operate and makes

12. J. R. Durig and D. J. Antion (to be published).
13. J. R. Durig and F. G. Baglin (to be published).

272   SECTION 5

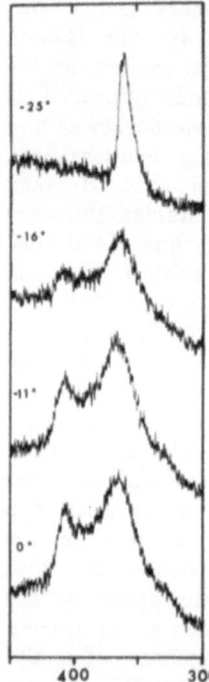

Fig. 4. Portion of Raman spectrum
of oxalyl bromide showing temperature dependence of 407-cm⁻¹ band.

available vibrational spectra over a large temperature range, a fact which should more than compen-
sate for the small reduction of signal.

ACKNOWLEDGMENTS

The authors wish to thank Thomas Giannaros of
the Custom Cutter Grind Company for constructing
the cradle. We should also like to thank the National Science Foundation for support under grant
number GP-8429, and for the purchase of the laser
with Grant Number GP-7079.

Raman Cell for Corrosive Liquids 5.3

L. S. Arighi* and M. V. Evans†

*Department of Chemistry, University of Wisconsin,
Madison, Wisconsin 53706*

This paper describes a Raman cell which has proved to be satisfactory for studying extremely reactive substances—for example, iodine pentafluoride and iodine heptafluoride. Other workers who have studied samples which are reactive to glass or quartz have relied upon the use of conventional cylindrical tubes made of more resistant glasses, such as aluminum phosphate[1] or translucent thin-wall tubes of Kel–F.[2] We have constructed a cell, which is shown in partially exploded view in Fig. 1.

This cell is made of Monel-bar stock assembled with silver solder; it has an 8-mm i.d. and an illuminated length of approximately 40 mm. The windows are made of 1.6-mm-thick sapphire; the larger entrance window is 12×50 mm; and the bottom exit window is 12 mm diam. These windows are gasketed to the ground surfaces of the cell by means of a 0.25-mm Teflon sheet which is coated with a very light film of Kel–F grease prior to assembly of the cell. We have found that this grease coating is essential if a leak-tight seal is to be maintained for extended periods of time. The windows are compressed against the gaskets by means of pressure plates, and the inevitable inequalities in the forces on the windows are

*Present address: Department of Chemistry, University of Portland, Portland, Oregon.

†Present address: Department of Chemistry, Worcester Polytechnic Institute, Worcester, Mass.

1. R. C. Lord, M. A. Lynch, W. C. Shumb, and E. J. Slowinski, J. Am. Chem. Soc. **72**, 522–527 (1950).
2. H. H. Hyman, L. A. Quarterman, M. Kilpatrick, and J. J. Katz, J. Phys. Chem. **65**, 123–127 (1961).

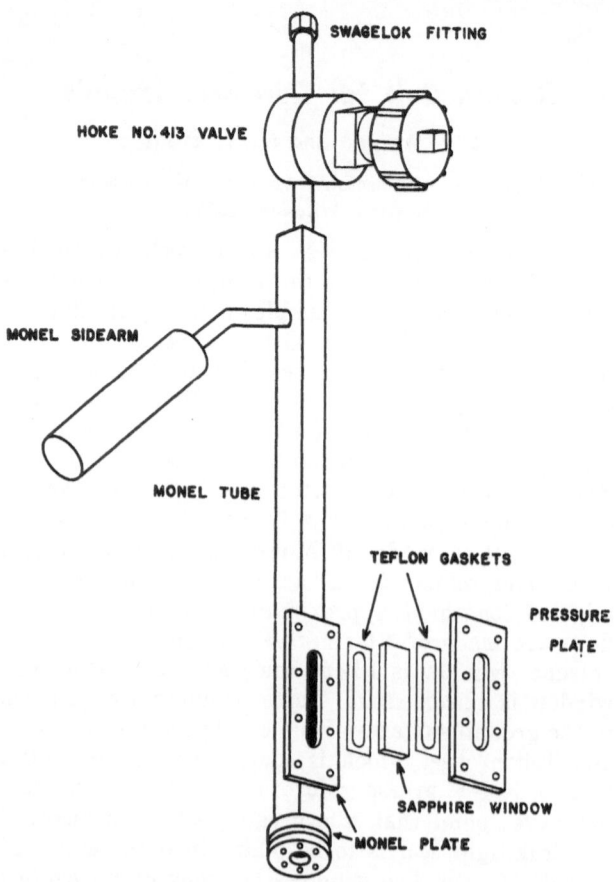

FIG. 1. Partially exploded view of the cell.

distributed by means of a Teflon cushion of approximately the same dimensions as those of the window-to-cell gasket. In the case of the exit window, it is essential that the Teflon gasket and cushion be prevented from sending scattered light into the spectrograph. This is accomplished by using a 0.1-mm-thick black polyethylene washer between the window and the Teflon cushion. Thus, the cell provides reasonably efficient illumination of the sample, yet minimizes direct scattering of the incident light into the spectrograph.

When working with highly reactive compounds, it is necessary to be able to remove the inevitable traces of suspended, solid products of reaction of the sample with residual water and other contaminants adsorbed on the cell walls. This was done by equipping the cell with a sidearm and bulb (also made of Monel), with a volume about double that of the cell itself, into which the liquid contents of the sealed-off cell could be poured. The window end of the cell could then be cooled by immersion in a dry-ice–trichloroethylene bath to distill the sample from the side arm into the cell proper, in order to remove nonvolatile contaminants. This arrangement has proved to be satisfactory in cleaning up residual, suspended impurities from many samples, and the repeated temperature and pressure cycling has not caused any detectable leaks in the cell.

The cell is not only chemically inert and resistant to temperature cycling but can withstand internal pressures of at least three atmospheres for periods of weeks. Of course, these advantages are not attained without sacrifice; the illuminated volume of the cell is at best approximately 1.8 ml, and the volume of liquid required to raise the meniscus comfortably above the illuminated space is nearly 3 ml. Also, reflection from the machined metal walls can cause appreciable error in depolarization measurements unless corrections are made for this diffusely reflected component of the incident light. In spite of these disadvantages, we have found that this cell performs very well when subjected to chemical abuse; yet it gives Raman spectra that are reasonably free of background and have usefully high Raman intensities.

5.4 A Low-Temperature Cell for Laser Raman Spectroscopy on Condensed Films and Nonvolatile Solids

D. F. Shriver, B. Swanson, and N. Nelson

*Department of Chemistry and Materials Research Center,
Northwestern University, Evanston, Illinois 60201*

The low-temperature cell described here is designed for laser Raman instruments (such as the Spex Ramalog) with 90° viewing and the beam entering the sample chamber from the bottom.[1] It yields Raman spectra under conditions which are analogous to those of the popular Wagner–Hornig low temperature infrared cell.[2] Thus, full advantage may be taken of the complementary nature of Raman and infrared spectra.

A currently used version of the cell is illustrated in Fig. 1, which is largely self-explanatory. The cell is attached to a vacuum system through an O–joint and after evacuation and introduction of refrigerant the sample vapor is directed onto the cold-finger. After a significant layer has been deposited, the Dewar and associated cold-finger is rotated by 180° so that the oblique cold-finger surface now faces the exit window rather than the sample inlet tube. In this configuration the laser beam enters the bottom window, impinges on the oblique surface of the cold-finger, and is scattered at 90° through the smaller window to the collection optics.

1. A more complex cell suitable for coaxial viewing has been described by G. L. Carlson, Spectrochim. Acta **24A**, 1519 (1968); and D. J. Antion and J. Durig, Appl. Spectry. **22**, 675 (1968); and a low temperature solution cell for 90° viewing is described by J. E. Griffiths, Appl. Spectry. **22**, 472 (1968).
2. E. L. Wagner and D. F. Hornig, J. Chem. Phys. **18**, 296 (1950).

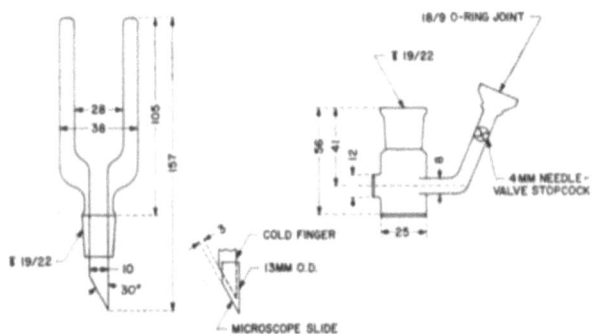

FIG. 1. The Dewar and associated cold-finger are depicted on the left. On the right is the cell body with window for entering beam (on the bottom) and scattered beam (on the side). The detail of the cold-finger tip (bottom middle drawing) illustrates a pocket for powdered solids, constructed from a 13-mm tube and microscope slide and held in place with room temperature vulcanizing silicone rubber cement. All dimensions are in mm.

Several details of construction bear special mention. The use of an O-ring joint on the inlet and a Teflon-in-glass needle–valve stopcock, avoids the exposure of incoming gases to stopcock grease. Thus, highly reactive substances and compounds of low volatility are handled with ease. To avoid distortion of the windows by conventional glassblowing techniques, the circular Pyrex flats are attached to the cell body by means of epoxy cement. The oblique surface on the cold finger is achieved by flattening the heated tube on one side and then grinding and polishing this surface to a uniform flatness. The 30° angle indicated in the drawing is employed rather than 45° to reduce specular reflection of the incident beam into the spectrometer. Furthermore, a small angle is advantageous to lengthen the image of the laser beam. Finally, we have found it advantageous to silver the well of the Dewar. This reduces heat loss and spurious signals due to bumping of liquid nitrogen. Also, some improvement in signal is achieved, presumably due to a second pass of the incident beam through the sample.

Among the modifications of this cell which we have used is a design with a Kovar-to-glass seal in the cold-finger and a polished copper tip brazed to the end. This cell is comparable to the all-glass version; however, occasional failure of the Kovar-to-glass seal may be experienced below $-78°C$.[3] Also, a three-pronged wire clip has been used to hold a pressed

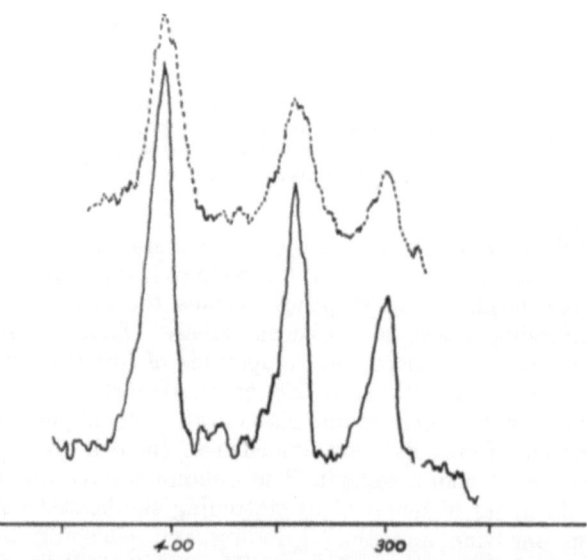

FIG. 2. Raman spectrum of $CD_3-C\equiv N-BF_3$ in the 250–450 cm^{-1} range; —— spectrum at $-196°C$, entrance slit 150 μ, period 20 sec, gain 5000, speed 45 cm^{-1}/min; – – – spectrum at 25°C, entrance slit 150 μ, period 20 sec, gain 7000, speed 27 cm^{-1}/min. A Spex 1400-II double monochromator was employed with a 6328 Å He–Ne laser and photon counting detector.

3. D. F. Shriver, *The Manipulation of Air-Sensitive Compounds* (McGraw–Hill Book Co., New York, 1969), p. 252.

pellet to the oblique surface of the cold-finger. This allows the observation of nonvolatile solids at low temperatures. Powdered solids which cannot be pressed successfully into pellets may be placed in a specially constructed pocket on the Dewar tip (see Fig. 1).

The performance of the cell on a sample of $CD_3-C\equiv N-BF_3$ is illustrated in Fig. 2, where it will be noted that the spectrum in the 250–450 cm^{-1} region is significantly sharper at $-196°C$ than at $25°C$. The sample was annealed at room temperature for one half hour before recooling to $-196°C$ and running the spectrum. To eliminate the condensation of moisture and freezing of the standard taper joint, a stream of warm air from a heat gun was directed on the cell during the scan.

ACKNOWLEDGMENTS

This cell has been used in research sponsored by the National Science Foundation, the Advanced Research Projects Agency, and the Alfred P. Sloan Foundation.

Low Temperature Sample Cell for a Laser 5.5
Excited Raman Spectrophotometer

John B. Bates* and William H. Smith†

Department of Chemistry, University of Kentucky, Lexington, Kentucky 40506

A cryostat for obtaining the Raman spectra of liquid and solid organic compounds was designed

*Present Address: Department of Chemistry, University of Maryland, College Park, Maryland 20742.
†Present Address: Department of Astrophysical Sciences, Princeton University, Princeton, New Jersey 08540.

for use on a Perkin–Elmer model LR-1 spectrophotometer using a helium-neon laser as the exciting
source.[1] The sample cell for the cryostat was designed
to minimize the amount of sample required for obtaining good spectra. By focusing the laser beam, it
was possible to reduce the beam to a size allowing

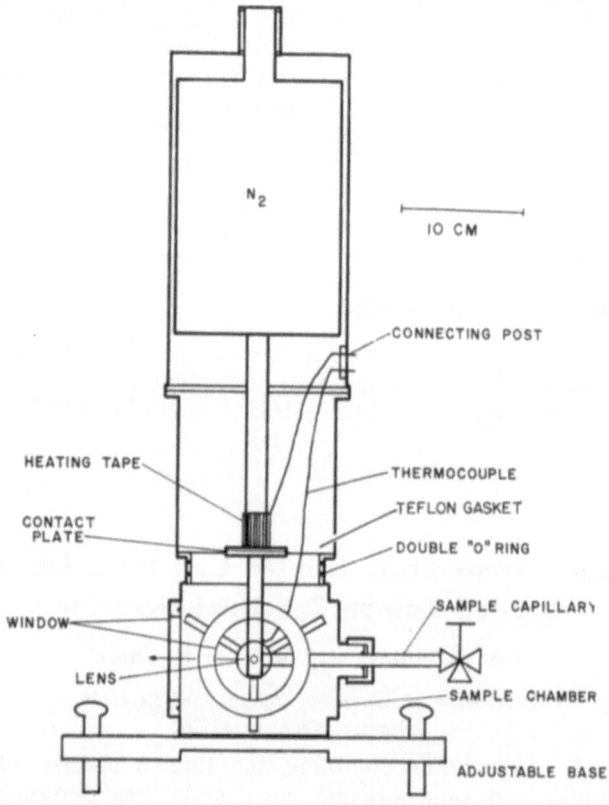

FIG. 1. Schematic diagram of the cryostat as viewed from
the laser.

1. H. A. Szymanski, *Raman Spectroscopy* (Plenum Press,
Inc., New York, 1967), p. 44.

samples as small as 0.15 ml of liquid to be used.
Since such a small sample acted as a point source
for the forward scattered light, it was necessary to
construct a Pyrex sample cell which would act as a
lens to focus the divergent light rays onto the instrument slits.

Figure 1 shows a schematic diagram of the cell.
The outer brass can is connected to a pumping station through a flexible brass hose. The inner liquid
nitrogen can is made of stainless steel and terminates
in a hollow stainless rod and brass plate. Heating
tape attached near the base of the rod with leads
attached to the terminal posts provides a means of
maintaining a constant temperature above 85 K as
well as for rapid heating of the rod and sample cell.
A solid copper rod is attached to the brass plate at
the base of the Dewar and extends into the sample
chamber where it is clamped around the Pyrex
sample tube. The thermocouple is attached to the
adjustment bolt on the rod (Fig. 2) and connected
to external leads through the terminal posts.

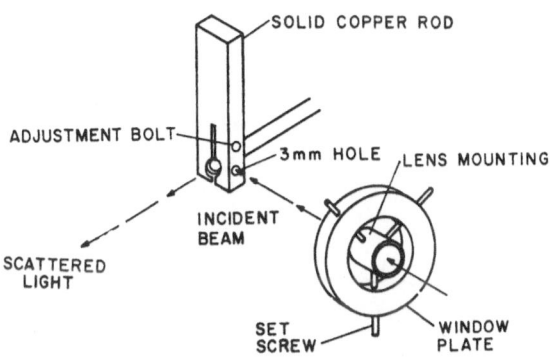

FIG. 2. Details of the interior of a sample chamber and lens
mounting assembly.

The metal Dewar is joined to the sample chamber by an additional brass section. The joint at the sample chamber is a double ''O'' ring seal with the upper portion of the cryostat resting on a Teflon gasket. This arrangement permits a rotation of the cell at this point so that the sample cell and copper rod can be aligned properly. The entire cell is bolted to an aluminum plate fitted with three large set screws by means of which the cell can be aligned with the laser and the monochromator slits. The Pyrex windows are held against ''O'' ring seals by brass ring gaskets. The window clamp on the laser side was made in such a way to contain an aluminum lens mounting (Fig. 2). Three set screws in the clamp permit a wide adjustment range for centering the laser beam onto any point within the sample bulb.

The Pyrex sample tube is shown in detail in Fig. 3. The capillary is 10-mm o.d. with a 2-mm bore. One end was extruded to form the sample bulb having a total volume of about 0.3 mliter. The bulb was shaped

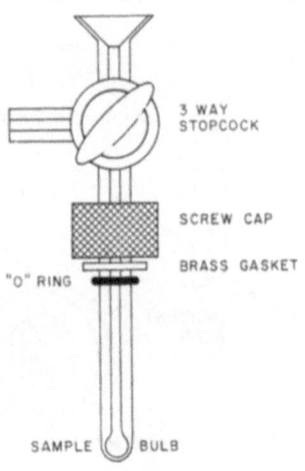

FIG. 3. Pyrex sample bulb.

so as to produce a lens effect which condenses the forward scattered light onto the monochromator slit. The open end of the capillary extends outside the sampler chamber and terminates in a 3-way stopcock. One arm of the stopcock is fitted with a ball socket, and the other arm provides a connection to a vacuum line so that the sample bulb may be evacuated prior to filling with a sample. The capillary is joined to the sample chamber by a threaded extension (Fig. 1) which is beveled at one end to form an ''O'' ring groove. The brass gasket (Fig. 3) is tightened against the O ring by a threaded brass cap thus forming a vacuum tight seal. Prior to tightening the cap, the capillary has sufficient freedom to allow for proper alignment of the bulb in the sample chamber.

Details of the sample chamber are shown in Fig. 2. The sample bulb is positioned in the copper rod so that it is exposed to the laser beam through a 3-mm hole in the rod. Good thermal contact is achieved by tightening the lower split end of the rod around the bulb. The lens mounting assembly consists of a small plano-convex lens (2.2-cm diam) with a 3.5-cm focal length secured in an aluminum cylinder. The cylinder is placed directly on the Pyrex window and can be adjusted with the three set screws.

The Raman instrument for which this cell was specifically designed is shown in Figs. 10, 11, 50, and 51 of Ref. 1. The assembled cell was placed on the sample bench of the LR-1 after removing the top cover of the sampling area (Fig. 51 of Ref. 1). After distilling a sample into the sample bulb, it was necessary to make several adjustments in order to align the cell with the beam and the monochromator slits. With some samples it was necessary to mount the laser independently of the instrument and adjust the laser as well as the cell. After making gross adjustments, a maximum intensity of the scattered

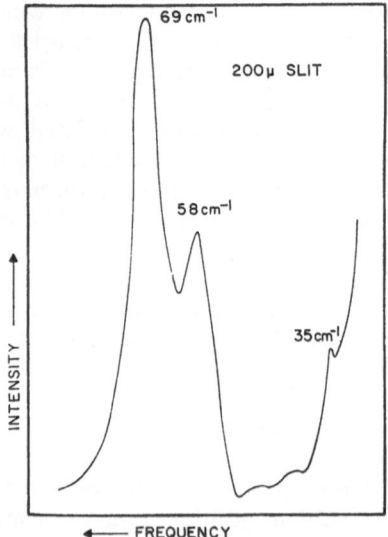

FIG. 4. Raman spectrum in the lattice region of polycrystalline C_3S_2 at 85 K.

light could be achieved by focusing the laser on various sections of the sample. By covering the back of the capillary and sample chamber with black felt, it was possible to reduce the background scattering by as much as 10%.

In Figs. 4 and 5 are displayed two examples from the Raman spectrum of carbon subsulfide at 85 K.[2] The sample used in this experiment occupied less than one-half of the sample bulb (about 0.15 mliter). The cell was aligned by maximizing the scattering from the band at 69 cm^{-1}. This region of the spectrum was scanned with a 5-cm^{-1} spectral slitwidth (200-μ mechanical slitwidth), and the region around 485

2. The complete infrared and Raman spectra of C_3S_2 will be discussed in a forthcoming publication.

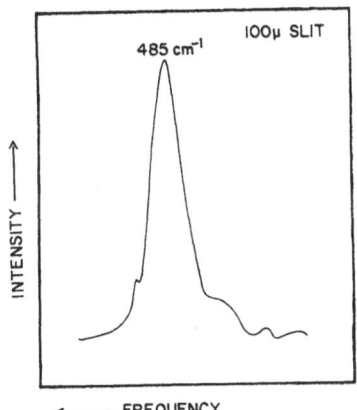

FIG. 5. Raman spectrum of the ν_2 (Σ_g^+) internal mode of polycrystalline C_3S_2 at 85 K.

cm^{-1} was scanned with a 2 cm^{-1} spectral slitwidth (100-μ mechanical slitwidth). These spectra were excited using a 10 mW He–Ne laser.

These two examples are typical of the spectra that can be obtained with small samples at low temperatures. By varying the voltage applied to the heating tape and regulating the amount of coolant (liquid N_2, dry-ice, acetone, etc.) in the Dewar, it is possible to maintain constant temperatures from 85–293 K so that the spectrum of the liquid as well as the effect of temperature changes on the spectrum of the crystal can be studied in a single experimental set up. In addition, our Raman cold cell was built so that it could be easily converted into a low temperature infrared cell.

5.6   A Laser-Raman Cell for Pressurized Liquids

R. Cavagnat, J. J. Martin, and G. Turrell

Laboratoire de Spectroscopie Infrarouge, associé au C.N.R.S.,
Faculté des Sciences de Bordeaux, 33—Talence, France

In recent years the development of the laser source
has greatly extended the range of application of
Raman spectroscopy. Now, as in the pre-laser era,
most Raman studies of ionic solutions are carried out
using water as the solvent. However, in many cases the
use of nonaqueous solvents such as NH_3, SO_2, HCl,
etc., is dictated by the nature of the solute. Furthermore, interest in fundamental properties of such nonaqueous solvent systems makes even more desirable
the development of experimental techniques suitable
for the investigation of pressurized liquids.

In the present note, a cell is described which has
been specifically designed for Raman studies of
pressurized nonaqueous solutions. Analogous work on
the ir spectra of such systems has been carried out
using a high-pressure ir cell which was developed
several years ago in this laboratory.[1] Thus, the availability of the cell described here allows Raman
spectroscopy to play its traditional complementary
role to the ir in the analyses of the vibrational spectra
of complex solute ions, and in investigations of solute–
solvent interactions in nonaqueous solutions.

The Raman cell with its cold trap and connection
tubulation is shown in Fig. 1. A detail of the cell
assembly is shown in Fig. 2. The solution is contained
in the quartz tube (A), which is fitted at either end
with quartz windows (B). Laser excitation enters
through these windows, making multiple passes along
the axis of the cell. The tube is supported by a stain-

1. E. Fishman, Appl. Opt. 1, 493 (1962).

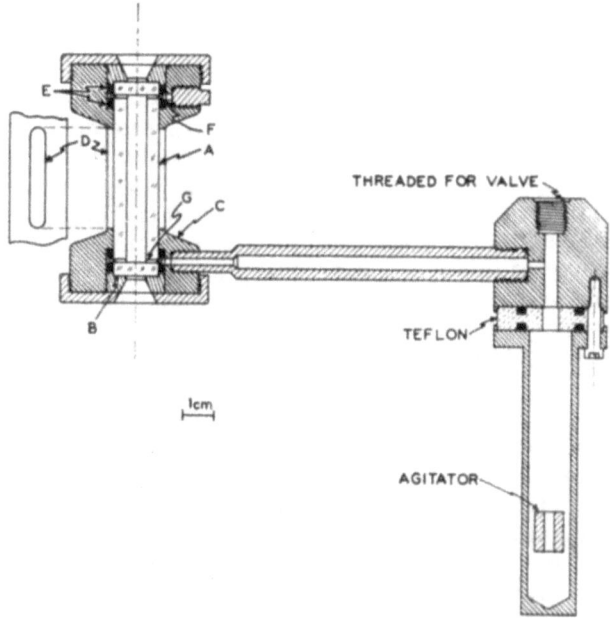

FIG. 1. Raman cell and trap.

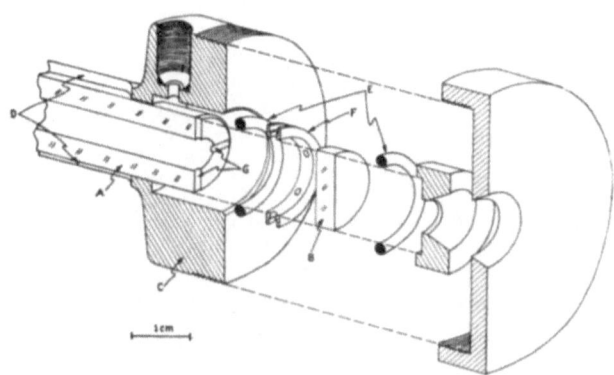

FIG. 2. Raman cell assembly.

less steel body (C), which is slotted at either side (D) to allow exit of the scattered light.

Pressure seals are provided at each end of the cell by pairs of silicon–rubber O-rings (E) separated by pierced stainless steel rings (F), as shown in the detail of Fig. 2. The purpose of the steel rings is to provide a path for entrance of the liquid without sacrificing the mechanical strength offered by the cylindrical symmetry of the quartz tube. Small grooves (G) ground in the ends of the tube allow the liquid to pass the windows and enter the cell.

The cell is easily filled, after initial evacuation of the system, by condensing a known volume of gas (e.g., NH_3, which is to serve as the solvent) in the trap using liquid nitrogen as the coolant. The desired quantity of solute can be placed in the trap before assembly, or, if it is sufficiently volatile, distilled into the trap after evacuation. When condensation is complete, the valve is closed and the trap is allowed to return to room temperature. The entire assembly can be removed

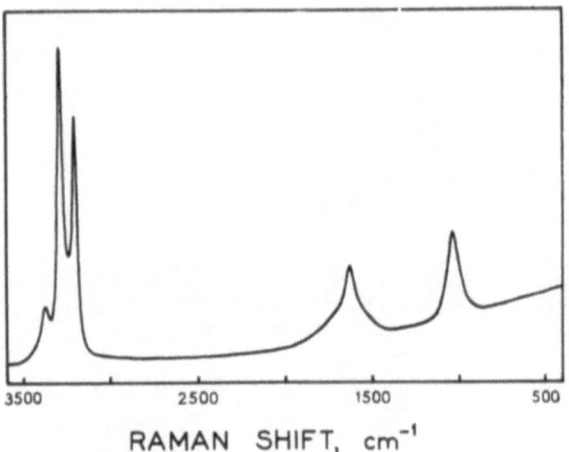

RAMAN   SHIFT,   cm^{-1}

FIG. 3. Raman spectrum of liquid NH_3 at room temperature.

from the vacuum system and agitated to aid dissolution of the solute. The solution is rapidly transferred from the trap to the cell by heating the former for about a minute with a hair dryer. Care should be taken to force all bubbles from the cell, as they produce large amounts of anamolous light scattering.

The cell is mounted in a transfer-optical system similar to that used with the standard liquid cell in the Coderg model CH 1 Raman spectrograph. A multipass attachment in the exciting beam provides a several-fold increase in Raman intensity.

The spectrum of pure liquid ammonia at room temperature obtained using the present cell in the Coderg instrument equipped with a O. I. P. 150 MW laser, is shown in Fig. 3. This spectrum is of significantly higher quality than those reported earlier for liquid ammonia under similar conditions.[2,3]

In the N–H stretching region the intensity pattern of the three bands at 3382 cm^{-1} (ν_3) and 3215 and 3300 cm^{-1} (Fermi doublet ν_1, $2\nu_4$) is characteristic of ammonia at room temperature. At low temperatures the outer lines of this triplet become relatively more intense.[4,5] The spectrum of Fig. 3 in the hydrogen-stretching region is similar to that reported recently by Ceccaldi and Leicknam, who did not, however, detect other vibrational fundamentals of liquid ammonia.[6] In the present work, bands of medium intensity at 1645 cm^{-1} (ν_4) and 1047 cm^{-1} (ν_2) were observed. Contrary to the suggestion of Kinumaki and Aida,[3] the latter band does not appear to be a doublet.

2. S. Bhagavantam, Indian J. Phys. **5**, 54 (1930).
3. S. Kinumaki and K. Aida, Sci. Rept. Res. Inst., Tohoku Univ. Ser. **A6**, 186 (1954).
4. P. Daure, Ann. Phys. **12**, 375 (1929).
5. C. A. Plint, R. M. B. Small, and H. L. Welsh, Can. J. Phys. **3**, 3653 (1954).
6. M. Ceccaldi and J. P. Leicknam, C. E. N. Rept. No. 6824, 14, February (1968).

Preliminary studies of the Raman spectra of small polyatomic ions in liquid ammonia solution have been carried out using the cell described here. The spectrum of the cyanate ion exhibits Raman lines at 2149 cm^{-1} (ν_3) and the Fermi doublet ν_1, $2\nu_2$ at 1206 and 1294 cm^{-1}. These frequencies are to be compared with the corresponding values of 2171 cm^{-1} (ν_3), 1225 and 1315 cm^{-1} ($\nu_1, 2\nu_2$) reported by Cleveland for the aqueous solution.[7] Thus the bond-stretching force constants of NCO$^-$ are substantially larger for the ion in water solution than in ammonia, an effect which is probably due primarily to the relatively higher electron-donor ability of water compared to NH$_3$.

5.7   Use of Commercial Ultraviolet Stabilizers as Optical-Filter Solutions in Raman Spectroscopy

J. M. Brown, A. C. Chapman, and D. A. E. Rendell

Research Department, Albright & Wilson, Ltd. (Oldbury Division), Oldbury, Warley, Worcestershire, England

In the conventional Raman source using Hg 4358-Å excitation, it is necessary to place an optical filter between the lamp and the sample to remove radiation in the violet and near uv regions of the spectrum which might decompose the sample. At one time, a saturated solution of sodium nitrite was almost universally employed for this purpose but this has now been largely superseded by solutions of aromatic nitrocompounds such as *m*-dinitrobenzene and the nitrotoluenes. Although these are more efficient filters than sodium nitrite, they suffer from a disadvantage; their transmission at 4358 Å falls off markedly after

7. F. F. Cleveland, J. Am. Chem. Soc. **63**, 622 (1941).

a few hours irradiation owing to photodecomposition. Frequent renewal is necessary.

A number of compounds which are employed commercially as ultraviolet filters and stabilizers for protecting plastics articles from the damaging effects of ultraviolet light, have been tested in this application as possible replacements for the nitrotoluenes. One of these compounds, 2,2′,4-hydroxy, 4′-methoxybenzophenone, sold commercially under the trade name of Uvinul 490 by the General Aniline and Film Corporation (Fine Dyestuffs & Chemicals Ltd. in the U.K.), has been found to give a filter system of equivalent performance to that given by the aromatic nitrocompound, with one difference; Uvinul is stable under Toronto-arc irradiation for a period of several weeks.

In Raman spectrometers, using photoelectric detection, it is also desirable to remove the green and yellow mercury lines from the lamp spectrum to reduce the background due to scattered radiation in the monochromator. The addition of methyl violet to the filter solution is effective in this respect, although the mixed solution is even less stable under irradiation. Methyl violet may similarly be added to the Uvinul 490 solution but without impairing its stability.

The absorption curves of filter solutions based on nitrotoluene and Uvinul 490 are shown in Fig. 1. The solutions were made up as follows: (a) p-nitrotoluene (15 g) plus methyl violet (0.25 g) dissolved in ethanol and made up to 100 ml; (b) Uvinul 490 (1 g) plus methyl violet (0.25 g) dissolved in ethanol and made up to 100 ml.

These curves refer to a path length of 10 mm, though in practice, a filter jacket of a thickness of about 4 mm is used. In use (based on measurements of the intensities of Raman lines with and without the filter solution), solution (a) transmits about 80%

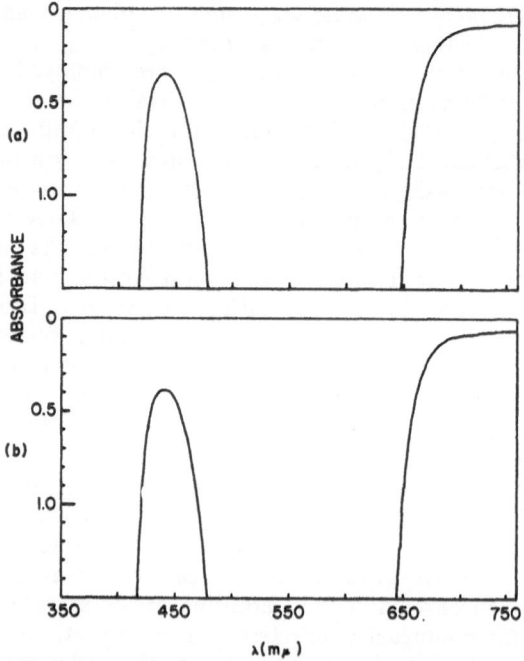

Fig. 1. Absorption curves of filter solutions based on (a) nitrotoluene and (b) Uvinul 490.

of the 4358-Å line when first made up with freshly recrystallized *p*-nitrotoluene. The comparative figure for solution (b) is about 60%. The performance of solution (a) however, becomes inferior to that of solution (b) after some 4-h irradiation and continues to deteriorate.

Apart from its greater stability, the use of Uvinul 490 avoids the toxic hazards associated with nitrocompounds. Furthermore, the solution is much less concentrated than the *p*-nitrotoluene solution which is almost saturated. This is particularly advantageous in systems where the filter solution is circulated.

Signal Enhancement Device for Use with Small Area Photocathodes in Raman Spectroscopy

5.8

James E. Griffiths

Bell Telephone Laboratories, Inc., Murray Hill, New Jersey

During the past few years significant advances have been made in reducing various noise levels in laser Raman spectroscopy. Particularly striking improvements include the reduction in stray light by using double monochromators,[1-3] decreased ghost intensities by interferometrically controlling the grating ruling engines, improved detection methods [4-6] and substantial reduction in dark noise by appropriate tube design,[7] and by using small area photocathode surfaces.[7, 8] In this connection we now use an ITT FW130 photomultiplier tube equipped with a slit-shaped (1.0×0.1 mm) S-20 photocathode mounted in a TE-104 Products for Research Inc. thermoelectrically coolable (to $-25°C$) tube holder. In this chamber, however, the cathode surface is about 2.6 in. from the exit slit and the bulk of the

1. Jarrell–Ash Co., Preliminary Tech. Release, EB-138, 1967.
2. R. C. Hawes, K. P. George, D. C. Nelson, and R. Beckwith, Anal. Chem. **38**, 1842 (1966).
3. D. Landon and S. P. S. Porto, Appl. Opt. **4**, 762 (1965).
4. Yoh-Han Pao, R. N. Zitter, and J. E. Griffiths, J. Opt. Soc. Am. **56**, 1133 (1966).
5. Yoh-Han Pao and J. E. Griffiths, J. Chem. Phys. **46**, 1671 (1967).
6. J. E. Griffiths, Appl. Spectry. **22**, 469 (1968).
7. E. H. Eberhardt, in a series of Applications Notes available from ITT Industrial Laboratories, considers in detail questions relating to tube design and related noise sources.
8. J. Sharpe, EMI Electronics Ltd., Hayes, Middlesex, England, Document No. CP 5475 (1964) discusses ''Dark Current in Photomultiplier Tubes.''

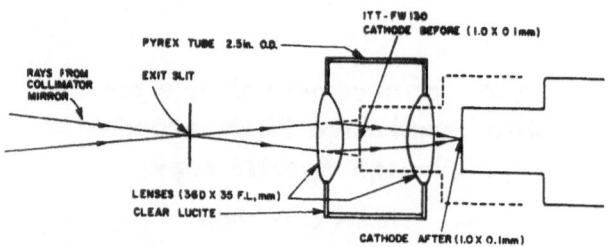

Fig. 1. A signal enhancement device for use with small area photocathodes,

rays diverging from the exit slit fail to impinge upon the photocathode and are irretrievably lost.

In order to eliminate this problem and still retain the insulating window, which prevents water condensation on the cooled photomultiplier tube parts, a simple and very inexpensive lens system has been incorporated. This is shown schematically in Fig. 1. In effect, a 2.5-in. diam Pyrex tube was cut and the ends were closed with 2.5-in. diam (0.125-in. thick) Lucite plates which had been prebored to accept the lenses. Lenses (Edmund's Scientific Co.) and Lucite plates were fixed in place with Smooth-On PSR-501 elastic bonding agent which retains its resiliency at the temperatures reached in the unit. Just prior to cementing the last plate surface, the tube was well flushed with dry nitrogen gas.

The advantage of the system is self-evident from Fig. 1 where the dashed lines indicate the position of the photocathode relative to the exit slit and the diverging light rays before backing off the tube housing and introducing the optical condenser system. The solid lines indicate the final relative positions of the components. A resilient light-tight stand-off allows the cooling unit to be backed off from the monochromator mounting plate and its resiliency allows enough flexibility for focusing purposes. With this crude arrangement, signal-to-noise ratios are increased at least one order of magnitude. Ideally, antireflection coated achromatic lenses with minimum spherical aberrations should be used, but the modest expense in dollars ($\sim$\$2.00) and construction time (1 h) of the present unit has much to recommend it.

SECTION 6
ULTRAVIOLET AND VISIBLE SPECTROSCOPY

An Ultraviolet Ruler   6.1

Elliott J. Levi

*Department of Chemistry, University of Cincinnati,
Cincinnati, Ohio 45221*

The reading of wavelengths (or wave numbers) at which maxima occur in ultraviolet spectroscopic analysis on a Cary model 11 recording spectrophotometer can be greatly simplified by the use of an "ultraviolet ruler." The ruler obviates the necessity of having to count wavelengths on recorder paper in which the abscissa is not graduated and on which only a few reference wavelengths are known.

The ruler is constructed from 1×1.5 in. soft wood, beveled on one side to an angle of $30°$, and the beveled side is then graduated in any convenient unit by scribing lines with a ballpoint pen.

The ruler is used by lining up a reference wavelength, such as 400 mμ, with the known wavelength on the trace and the wavelengths of other peaks can then be determined. The ultraviolet ruler thus functions as a portable abscissa for ungraduated recorder paper. Refer to Fig. 1 in the text.

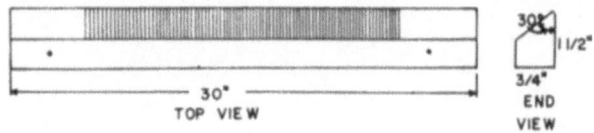

TOP VIEW

END VIEW

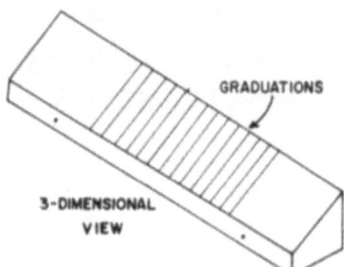

GRADUATIONS

3-DIMENSIONAL VIEW

FIG. 1. Ultraviolet Ruler.

6.2 Battery Eliminator for Spectral-Energy-Recording Adapter on Beckman DU Spectrophotometer

J. D. Chazin and D. L. Ohliger

St. Regis Paper Company, Technical Center, West Nyack, New York 10994

The Spectral-Energy-Recording Adapter (SERA)[1] has been an effective tool in this laboratory for qualitative as well as quantitative trace elemental analysis via flame photometric spectra. Originally a completely battery-powered instrument, the Beckman DU Spectrophotometer was simplified by the addition of the 73600 power supply, an ac-to-dc converter, which eliminated the need for all but one of the storage

1. Beckman Instruments, Inc., Fullerton, Calif.

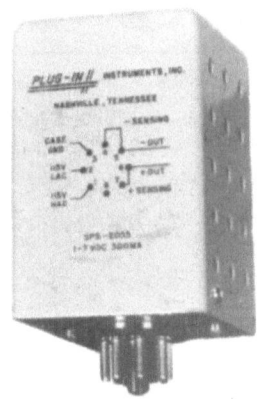

Fig. 1. "Plug-In" low-voltage-regulated dc power supply, Model No. SPS-2055-P.

and dry-cell batteries. However, the SERA still required the use of a 1.5-V type A (lantern) battery.

The function of the 1.5-V A battery is to energize the filament of the 2531 amplifier tube in the photomultiplier circuit when the SERA is in the RECORD position. This power source must be stable and noise free. The use of the A cell has created many problems and inconveniences: constant drift as the battery deteriorates with use and time; replacement of the battery every few months; limited shelf life of the batteries; etc.

The need for a stable, noise-free, constant-power source to replace the A cell has long been recognized. To resolve this problem, we have successfully adapted a "plug-in" low-voltage, regulated dc power supply[2] (Fig. 1) to the SERA. This source is an ac-to-dc converter with a 105–125-V-ac, 50–400-cps input and a regulated, adjustable output of 1–6.5 V at 0–300 mA dc. It is a solid-state miniature unit with electronic overload protection. Silicon transistors provide the desired stability and noise-free operation.

2. Plug-In Instruments, Inc., Nashville, Tenn.

Table I. Specifications for "plug-in" dc power supply.

Model No.	SPS—2055 P
Size	D (2$\frac{1}{4}$ in. width $\times$ 2$\frac{9}{32}$ in. diam $\times$ 3$\frac{1}{8}$ in. height)
Style	Plug-in, 8-pin octal
Input range (V, ac)	105–125 (50–400 cps power required)
Output range (V, dc)	1–6.5
Current (mA, dc)	0–300
Line regulation (mV, dc)	8
Load regulation (mV, dc)	5
Ripple (mV, per pulse)	1.5
Output impedance (Ω)	0.1 (floating output with shield between transformer primary and secondary)
Ambient temperature (°F)	20–130
Temperature coefficient (%/°F)	0.015 (nominal)

The specifications are listed in Table I and the circuit diagram is illustrated in Fig. 2.

Mounting of this unit is accomplished by simply modifying the 1.5-V battery housing on the SERA, as illustrated in Fig. 3. A hole, 1$\frac{3}{16}$ in. in diam is punched out of the base of the battery housing with a chassis punch. An eight-pin-octal tube socket with the input and output leads soldered in place is mounted with two 6-32 round-head brass machine screws. A 2-in.-diam cylindrical piece of Plexiglas, $\frac{1}{8}$ in. thick, is cut to 1 in. length. It is covered on one end with a piece of flat, $\frac{1}{8}$-in.-thick Plexiglas cut to the proper roundness and cemented in place. The cylinder is notched to allow access for the ac and dc leads. The Plexiglas case is held in place by a clamp

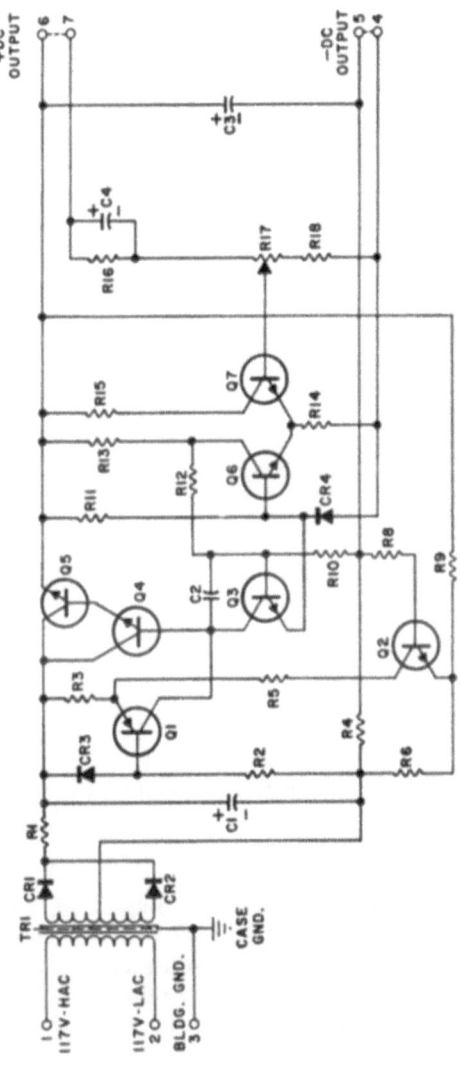

FIG. 2. Circuit diagram of "plug-in" dc power supply, type D.

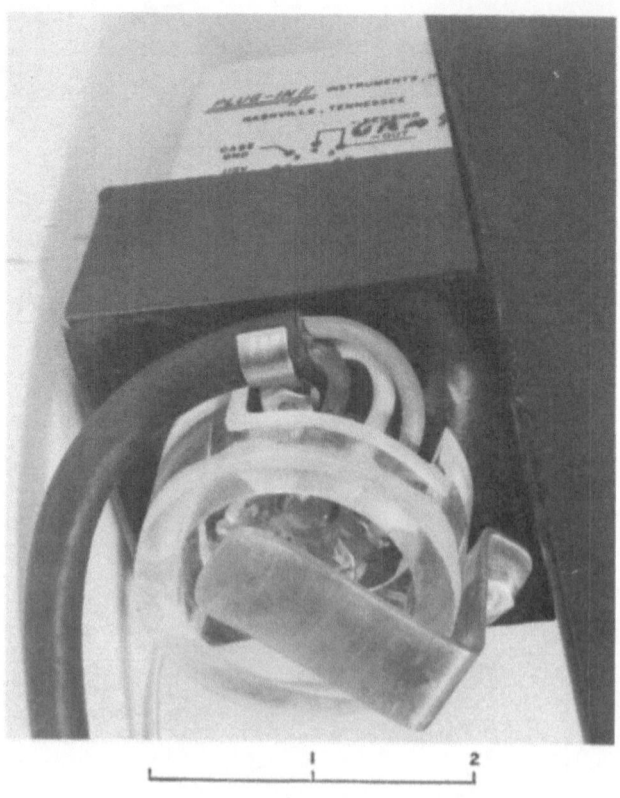

INCHES

FIG. 3. Modification of battery housing on SERA with "plug-in" dc power supply in place.

fabricated from $\frac{1}{16}$-in. brass stock, as illustrated, and mounted on the housing with a 6-32 brass machine screw and nut. This case serves as protection against accidental electrical shock or short circuiting.

The power supply is simply snapped into its socket, and the ac input lead with a three-prong wall plug is connected to the ac house current. A warmup period of 20 min is allowed the instrument in this setup, and the power supply is adjusted via a slotted knob recessed in the covering case of the plug-in unit. This knob operates a trimpot, which is adjusted to that voltage which will give a straight baseline on the recorder for the dark current (approximately 1.5 V). The SERA is then ready to operate.

We have been successfully operating this power supply in place of the battery for several months with excellent results. There has been no drift, and the noise level has not varied from that experienced with the 1.5-V A cell at all sensitivities.

Accurate Temperature Control of the   6.3
Beckman DK-2 Recording
Spectrophotometer

Jon M. Veigel

Frank J. Seiler Research Laboratory, Chemistry Division, Office of Aerospace Research, USAF Academy, Colorado 80840

The prime requirement for accurate kinetic analysis is that the temperature of the system being studied remain stable over the course of the reaction. As Benson has pointed out, typical reactions around room temperature will have a maximum accuracy in the rate constant of $\pm 0.1\%$ when the temperature is known to $\pm 0.01°C$.[1]

For those reactions that can be followed by changes in ultraviolet and visible spectra, temperature control must be effected for the solutions contained in the optical cells within a spectrophotometer. This is

---

1. S. W. Benson, *The Foundation of Chemical Kinetics*, (McGraw–Hill Book Company, Inc., New York, 1960) p. 92.

Fig. 1. Temperature regulator.

a particular problem for the Beckman DK-2 recording spectrophotometer since the cell compartment
normally serves as a part of a heat sink for the lamp.

However, very accurate temperature control can
be obtained by two modifications. A Beckman tap-
water-cooled lamp housing eliminates the lamp as a
heat source. The device pictured in Fig. 1 regulates
the cell temperature. The box fits the standard cell
compartment and holds the sample and reference 10-
cm cells in the same position as a standard cell holder.
The cells are in close contact with the tubes throughout their length. Excellent thermal contact is thus
maintained between the solutions being studied and
the water circulated through the hollow interior of

the compartment. The water enters through fittings on the box top while the notched flange provides room for the hoses which pass through the cutouts at each end. Rubber-gaskets ensure light-tightness. The cooling water is passed via short insulated tubing to a "Forma-Temp. Jr." thermostated circulator.

With this apparatus, temperature accuracy within the cells of $\pm 0.01°$–$0.02°C$ over a $15°$–$30°C$ temperature range were routinely maintained for one day or longer.

Attachment for Cary Model 14R   6.4
Spectrophotometer

Allen L. Olsen and Marian E. Hills

U. S. Naval Ordnance Test Station, China Lake, California 93555

Polarization studies with the Cary Model 14 spectrophotometer required the differences in absorbance units of the E vector in the vertical and horizontal positions for polarizing sheets as a function of wavelength. This presented no problem for the Cary Model 14R since it was equipped with a Datex system for converting spectral traces to digital data and recording on punched paper tape. Measurements of other Cary 14 instruments, not modified to accommodate the digital system, involve a laborious point-by-point treatment of the data to obtain these differences. This communication describes a simplified technique, applied to the Cary 14R-Datex system, to digitize spectral traces from other Cary 14 instruments. This procedure provides not only the differences in absorbance units for polarization studies but also file data of other spectra for reference uses.

FIG. 1. Knurled knob attached to the encoder shaft of the
Cary 14R spectrophotometer.

An attachment, a 3-in. knob with a knurled edge,
shown in Fig. 1, is secured to the encoder shaft of
the Cary 14R. Although segments of chart paper
from the Cary 14 may be used, it appeared advantageous to use the entire roll since this provides necessary tension on the chart paper. The pen and slit
control toggle switches are inactivated, and scan and
chart speeds are selected to correspond to the settings for the original trace. The pen, preferably without ink, is set at the specified wavelength and absorb-

ance values, the Datex system is set to record the initial data, and with activation of the master switch, the pen is made to follow the spectral trace by manually turning the knurled knob. In practice, the pen carriage is guided between the peak-to-peak noise level of the original trace. Removal of the knurled flange on the pen tip allowed the operator to see the advancing trace.

To evaluate the reproducibility of the procedure, measurements were made in triplicate of HN 22[1] polarizing sheet on the Cary 14R with a dynode setting of 3 and slit control 0. The scan was made from 8000 to 4000 Å, and readouts were made at 10-Å intervals. One of these spectral curves was reproduced by the manual technique. The distribution of mini–max absorbance units (AU) at readout points was determined for the two sets of triplicate runs. The estimate of standard deviation based on the mean range of triples[2] is 0.0014 and 0.0007 AU for the 400 triples in spectral runs and traces, respectively. These values are well within the reproducibility performance of the Cary instrument (± 0.002 AU in the 0 to 1 mode and ± 0.005 AU in the upper portion of the 1 to 2 mode).[3] It is not surprising that the trace technique indicates improvement over instrument reproducibility, since the noise level of the curve has been minimized.

Acknowledgments. The authors wish to acknowledge the valuable assistance of H. N. Browne, Jr., and E. A. Fay in the statistical treatment of the data.

1. Polaroid Corporation, Cambridge, Mass. 02139.

2. E. L. Crow, F. A. Davis, and M. W. Maxfield, ''Statistics Manual,'' U. S. Naval Ordnance Test Station, China Lake, Calif. 93555, NAVORD Rept. 3369 (1955), Table 12, p. 248.

3. *Instructions for Cary Recording Spectrophotometer Model 14R* (Cary Instruments, Monrovia, Calif.), R 1/64, p. 5.

SECTION 7
X-RAY SPECTROSCOPY

A Method of Producing Sturdy Specimens 7.1 of Pressed Powders for Use in X-ray Spectrochemical Analysis

Leonard Bean

National Bureau of Standards, Washington, D. C. 20234

Many investigators have been confronted with the problem of preparing sturdy pressed powder specimens for use in x-ray spectrochemical analysis. Some have ground or otherwise mixed the powdered material with a binder, such as boric acid[1] or cellulose powder prior to pressing in a mold. This dilutes the sample and causes some added uncertainty. One never knows whether the binder is evenly dispersed throughout the sample. Some investigators have pressed powders on a backing of boric acid,[1] cellulose powder,[2] or Boraxo.[3]

1. H. Rose, I. Adler, and F. J. Flanagan, "Use of La$_2$O as a Heavy Absorber in the X-ray Fluorescence Analysis of Silicate Rocks," U. S. Geological Survey Professional Paper 450-5 (1962), p. 80; Also see, "X-ray Fluorescence Analysis of the Light Elements in Rocks and Minerals," Appl. Spectry. **17**, 81 (1963).
2. G. Andermann and J. D. Allen, "X-ray Emission Analysis of Finished Cements," Anal. Chem. **33**, 1695 (1961).
3. T. M. Houseknecht and W. Patterson, "Sample Prepara-

In most of the techniques mentioned above, difficulties arise due to crumbling of the pressed specimen
around the edges. Some investigators have minimized
this effect by masking the surface to be irradiated and
spraying the edges with a plastic spray.[2]

Recently, in connection with the study of x-ray
spectrochemical methods in this laboratory, a technique has been developed which produces a sturdy
specimen with strong edges. No dilution of the powdered sample is necessary for the pressing operation.
In addition, the pressure can be applied or released
rapidly without shattering the specimen.

A regular mold made of tool steel is used. Three
additional accessories are prepared. One is a loose-
fitting steel sleeve, with sides approximately 0.16 cm
($\frac{1}{16}$ in.) thick, and outside diameter about 0.02 cm
(1/128 in.) less than the inside diameter of the
mold. The sleeve should be long enough to project
1.3 to 1.6 cm ($\frac{1}{2}$ to $\frac{5}{8}$ in.) above the mold when in use
so that it may be grasped with the fingers. The second
accessory is a steel plunger whose diameter is approximately 0.08 cm ($\frac{1}{32}$ in.) less than the inside diameter
of the sleeve and about 2.5 cm (1 in.) longer than
the sleeve. These two accessories are shown in Fig. 1.
The third accessory (not shown) is a piece of tungsten wire (or other stiff wire) approximately 0.076
cm (30 mil) in diameter and 5 cm (2 in.) long held
in a pin vise.

A specimen is prepared in the following manner:
The steel mold is placed in a position to receive a
sample. The mold may be fastened in a press as with
some motor-driven presses, or on a bench supported

tion for X-ray Analysis,'' Spectrographer's News Letter, 27,
No. 2 (1964), Publ. by Applied Research Laboratories,
Glendale, Calif.

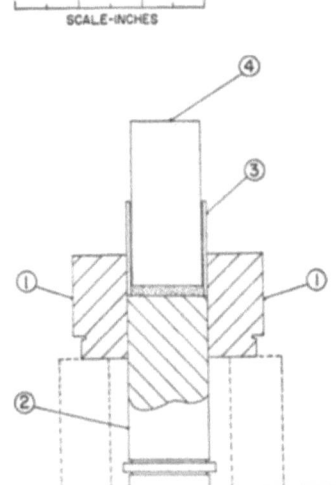

FIG. 1. Schematic side view of apparatus used for pressing specimens. 1. Main body of mold. 2. Lower press plunger 3. Steel sleeve. 4. Steel plunger for hand tamping sample.

by two blocks. In either case, the lower plunger is inserted part of the way into the mold. The sleeve is placed inside the mold, resting on top of the lower plunger. Approximately a 2-g sample (the amount is not critical) of the powder to be analyzed is poured inside the sleeve on top of the lower plunger. The powder is spread with the end of a spatula so that the top of the plunger is no longer visible. A spatula with its end ground off squarely (not rounded) is convenient for the purpose. The loose-fitting plunger is then used by hand to press *gently* and smooth the top of the powder. This is to insure that the bottom (the surface to be analyzed) will contain no voids. The end of the tungsten wire is then inserted and scraped around the inside of the sleeve with the end touching the top of the plunger. This procedure separates the powder from the sleeve so that the powder will not tend to be lifted when the sleeve is withdrawn. The sleeve is then lifted from the mold. (It is well to rotate the sleeve prior to removal.) With some powdered materials, it is unnecessary to use the loose-fitting plunger and stiff wire; spreading with the end of a spatula may be sufficient. The choice of whether or not to use the plunger will soon become apparent to the operator, based on the type of powder used.

This procedure leaves the powder on top of the lower plunger with a peripheral trench at the edge. A 5-ml beaker is filled with boric acid (crystal). The contents of the beaker is then poured on top of the powder to be analyzed, taking care to fill the trench first. The top plunger is placed in position and rotated by hand two or three times while in contact with the boric acid. The sample is then pressed to the desired pressure. With this technique, it is not necessary to apply or release the pressure slowly. (Rapid application and/or release of pressure are among the factors which can cause a specimen to shatter.[4] Some type of lubricant frequently obviates this difficulty. The boric acid edges seem to function as a lubricant between the powder and sides of the mold.) The specimen is extruded from the mold in the usual manner dictated by the type of press used. Suitable identification may be written on the boric acid backing with a ball point pen.

The specimen formed is quite sturdy. In Fig. 2 may be seen a photograph of the finished specimen as well as a cross section of a specimen. The edges of pressed boric acid (sometimes mixed with some of the powder) are quite strong and take the pressure exerted by most types of sample holders, so that the

---

4. W. D. Kingery, *Introduction to Ceramics* (John Wiley & Sons, Inc., New York, 1960), p. 42.

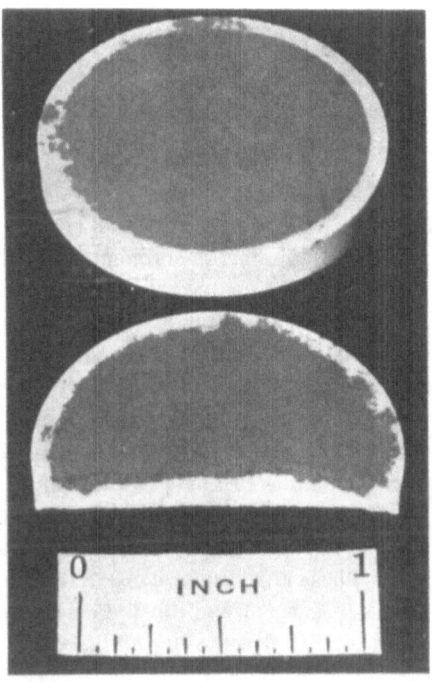

FIG. 2. Photograph of finished specimen and of specimen broken in two to show cross-sectional view.

powder being analyzed does not need to support this pressure. As a consequence, the edges (or corners) do not crumble and the specimen may be reused many times, if desired. Sturdy specimens have been prepared by this method at pressures from 350 to 4900 kg/cm² (5000 to 70 000 psi).

7.2   Sterotex—A Grinding Aid

Charles K. Matocha

Alcoa Research Laboratories, New Kensington, Pennsylvania

Powdered samples for x-ray analysis are often
ground in mills by using a rotating ring and puck
inside a grinding cylinder. Such commercial grinders
are the Disc Mill (Angstrom, Inc.), Bleuler Mill
(Applied Research Laboratories, Inc.), and Shatter-
box (Spex Industries, Inc.). A grinding aid is frequently added along with the sample to increase the
grinding efficiency, to minimize contamination caused
by abrasion with the grinding media, and to eliminate sample's sticking to the grinding surfaces and
thus aid in sample transfer. Grinding aids frequently
employed are soap powders, sodium alkylsulfonate,
and Boraxo. These products contain sodium; this
precludes their use when sodium is to be determined.
Sterotex powder,[1] a vegetable-fat product that is
free of the elements determined by x-ray, has been
tested and used as a grinding aid. Sterotex, a fine
powder about 95% −200 mesh, easily and uniformly
disperses with the sample. When required, it can be
premixed with the sample to minimize spot-impacting
of the sample on the grinding surfaces. To eliminate
this problem with, for example, Boraxo, the as-re-
ceived product should be ball-milled to about −50
mesh. Sterotex also imparts some binding action
when briquets are prepared from the pulverized
sample.

---

1. Available from Capital City Products Company, PO Box
   569, Columbus, Ohio 43216.

It is not implied that Sterotex is a universal grinding aid, just as the ring-and-puck mill is not a universal grinder. Some samples have been encountered that do not respond to this vigorous grinding action with any of the grinding aids that were tested. In general, for samples that grind satisfactorily with these mills, Sterotex can replace the grinding aids previously mentioned. A sample charge weight, weight of grinding aid, and grinding time can then be established for producing a pulverized sample suitable for x-ray analysis.

Preparation of Glass Blanks by Fusion   7.3
with Lithium Tetraborate for
X-Ray Spectrographic Analysis

C. H. Drummond

Owens-Illinois Glass Company, Technical Center, Toledo, Ohio

This sample-preparation technique is used for obtaining homogeneous, stress-free glass blanks for x-ray analysis. The glass blanks are ground to a 600 grit finish. Previously published papers have given the basic theories of the method.[1-6]

1. E. P. Bertin and R. J. Longobucco, Norelco Reporter 9, No. 2, 31 (1962).
2. T. J. Cullen, Spex Speaker 6, No. 4, 1 (1961).
3. F. Claisse, Quebec, Dep. Mines Prelim. Rept. 327, 24 pp. (1956).
4. H. T. Dryer, Advan. X-Ray Anal. 6, 447–458 (1962).
5. H. J. Rose, Jr., I. Adler, and F. J. Flanagan, Appl. Spectry. 17, 81–85 (1963).
6. E. E. Welday, A. K. Baird, D. B. McIntyre, and K. W. Madlem, Am. Mineralogist 49, 889 (1964).

The method consists of weighing one part of sample (3 g) to three parts of anhydrous lithium tetraborate (9 g) and transferring to a 50 ml platinum–gold[7] crucible with a lid. The mixture is stirred in the crucible with a platinum stirring rod and melted at 1250°C for 10 min.

After cooling to the touch on a zircon plate, the glass blank will fall from the inverted crucible. The sample is crushed to pea size in a mortar and pestle and ground for 1 min in a Pica mill with a tungsten carbide mortar and balls. The ground powder is transferred to the crucible, remelted at 1250°C for 5 min, and the crucible removed and cooled. The glass blank is removed and reinserted so that the *seeds* that were on the bottom are now on the top and melted at 1250°C for 5 min. The glass blank is now in a homogeneous and seed-free condition. The crucible is removed and transferred to an annealer held at 650°C. After $\frac{1}{2}$ h, the power is turned off and the annealer is allowed to cool at the rate of 1–1.5°C/min. This is most conveniently done overnight. After annealing, the glass blank is ground with 600 silicon carbide grit on a "Spitfire" 12 in. lapping machine.[8]

The method has also been used to prepare glass blanks that were one part sample, one part lanthanum oxide, and eight parts lithium tetraborate. Samples have included all types of glasses, refractories, sand, feldspar, and related glass batch materials. All samples have yielded homogeneous glass blanks that were capable of being annealed and ground without cracking.

7. Englehard Industries, Baker Platinum Division, Newark, N. J. Form No. 202, low form—wide bottom, 95% platinum–5% gold, of 0.020 in. thickness, with lid to fit.
8. Precision Machine Company, Chicago, Ill., model SP-ML-12.

Flat vs Curved Aluminum Planchets   7.4
for X-Ray Fluorescence Spectrometry

Harvey D. Spitz

Johnson & Johnson Research, New Brunswick,
New Jersey 08903

X-Ray analysis has become an increasingly important tool for the quantitative determination of trace quantities of elements in various samples.

It is surprising to find the small number of different sample containers that are employed in trace analysis. In our laboratory, flat aluminum planchets are often used as sample holders for the determination of a specific element by x-ray fluorescence analysis.

Usually, a solution containing the trace constituent is placed dropwise on the planchet and evaporated. It is difficult to keep the material near the center of the disk as the liquid randomly spreads to the edge of the planchet. This can cause large errors in trace analysis, because the instrument will "see" only that portion of the sample which is in the center of the planchet. In order to reduce this source of error, a curved planchet was prepared. It has the advantage of keeping the liquid (and therefore the sample) in the center portion of the disk. The flat planchets were prepared by punching out the disk (from an aluminum weighing dish) with a die punch of the correct diameter. The curved planchets were then prepared by placing a flat planchet in an evaporating dish, covering the disk with a folded tissue, and then curving it with a pestle by applying light pressure with a rotary motion. This particular equipment was chosen so as to give the desired curvature to the planchets.

A comparison was made between the curved and flat aluminum planchets. Zinc and chlorine (as polyvinyl chloride) were placed on the planchets (in triplicate) with 1.00 and 0.50 ml volumes of solution representing different levels of concentration. The solvent for zinc (actually $ZnSO_4 \cdot 7 H_2O$) was 80% ethanol, while 1,2 dichlorethane was used for the polyvinyl chloride. These elements were chosen so that the results obtained would be representative of both metallic and nonmetallic elements. The results are given in Table I.

The results seem to indicate that the curved planchets produce a slight loss in sensitivity. It should be noted that the variation found is greatest with the flat planchet containing zinc at the 9 μg range. This shows a slightly lower average result than one might expect according to the remaining data. From a practical point of view, one might conclude that there is no significant difference between the two types of planchets considering the low concentrations and the range in counts per second obtained.

Although the precision was comparable with both types of planchets, one would expect greater accuracy

Table I. Comparison of flat and curved planchets.

Type of planchet	Sample on planchet (μg)		Ave. count Sec[a]
Flat	4.6	(Zn)	188
Curved	4.6	(Zn)	157
Flat	9.2	(Zn)	330
Curved	9.2	(Zn)	340
Flat	500	(polyvinyl chloride)	338
Curved	500	(polyvinyl chloride)	298
Flat	1000	(polyvinyl chloride)	658
Curved	1000	(polyvinyl chloride)	617

[a] Corrected for blank.

and precision with curved planchets when large volumes of solution must be evaporated on the planchet. This technique of using curved aluminum planchets should be of great value when analyzing samples which form solid residues upon evaporation of the sample solvent.

Acknowledgment—The author wishes to thank J. P. Schelz and L. Bovino for helpful advice in preparing this note.

Rotating Sample Holder for X-Ray   7.5
Emission Spectrometry

N. P. Quebbeman and C. W. Jones

General Electric Company, St. Petersburg, Florida 33733

A rotating sample holder is invaluable in x-ray emission analysis of nonhomogeneous samples because when such samples are rotated,[1,2] different sample areas are observed continuously, thus permitting an averaging of irregularities. This feature of the sample holder is of significance in the nondestructive testing of materials in that it provides more reliable analytical data. The rotating sample holder described in this paper is designed and constructed for use on a General Electric XRD-6 x-ray emission spectrometer. The reduced x-ray intensity variation obtained by continuous sample rotation results in significant improvement in over-all precision compared with intensity variation obtained by rotation only between readings.

1. E. P. Bertin and R. J. Longobucco, Advan. X-Ray Anal. **5**, 447 (1962).

2. "Turntable Sample Holder for X-Ray Spectroscopy," The Spex Speaker, Metuchen, N. J., Spex Industries, Inc., Vol. 5, No. 2, p. 6 (June 1960).

FIG. 1. Components needed for rotation about an axis normal to the shield: sample drawer, motor, contra-angle U, assorted-size cups, and cup holders.

The basic components of the rotating sample holder are: a conventional sample drawer, Catalog No. A4961-H, used on General Electric XRD-6 x-ray emission spectrometers; a 30-rpm motor; and a contra-angle U gearing device, used on S.S. White dental drill handpieces. (Detailed drawings of the sample holder are available from the authors on request.)

The holder may be used to rotate samples about an axis either normal or parallel to the shield. Figure 1 shows the components required for rotation about an axis normal to the shield. When the sample holder is used for rotation about an axis parallel to the shield, the contra-angle U is oriented to the position shown in Fig. 2.

Comparative evaluation data was obtained for continuous rotation and rotation only between readings about an axis normal to the shield. The test specimen was a fairly uniform briquet made by press-

Fig. 2. A ceramic cylinder held in place on the platform by an aluminum arbor adaptor for rotation about an axis parallel to the shield.

ing a 5-g sample of National Bureau of Standards 37-D sheet brass into a 1-in. diam disk at 50,000 lb/in² (Ref. 3) for 2 min. Ten spectrometer readings were taken for continuous rotation and ten for rotation only between readings. Table I gives the spectrometer parameters and test conditions, and Table II gives comparative x-ray intensities for the two conditions. The counting error was minimized by taking a sufficiently large number of counts.

A similar comparison was made for sample rotation about an axis parallel to the shield. The test specimen was fairly nonuniform and consisted of a ceramic cylinder coated with molybdenum. Ten spectrometer readings were taken for continuous rotation and ten for rotation only between readings. Table I gives the spectrometer parameters and test conditions, and Table III gives comparative x-ray intensities for the two conditions. Again, the counting error was minimized by taking a sufficiently large numbr of counts.

3. T. C. Cullen, Anal. Chem. **33**, 1342 (1961).

Table I. Spectrometer parameters and conditions.

Component or condition	Brass-disk sample	Mo-coated ceramic sample
X-Ray tube	GE EA-75, Cr target	GE EA-75, W target
Power parameters	25 kV, 20 MA	50 kV, 50 MA
Analyzing crystal	Lithium fluoride	Lithium fluoride
Counter tube	SPG-7 gas-flow proportional	SPG-4 scintillation, mounted in rear position in preamplifier, front position unoccupied
Soller slit	0.020 in.	0.020 in.
Amplifier gain	Coarse setting, 16; fine setting, 90	Coarse setting, 8; fine setting, 45
Counting time	100 sec	100 sec
Sample analyzed	NBS 37-D sheet brass	Molybdenum-coated ceramic
Element	Copper	Molybdenum
Spectral line measured	Cu $K\alpha$, 1st Order	Mo $K\alpha$, 1st Order
E, base line voltage	2 V	5 V
E, pulse-height selector	6 V	Not used
High-voltage front selector potentiometer setting	860	730
High-voltage rear selector potentiometer setting	Off	Maximum

Table II. Comparison of the variation in x-ray intensity for pressed sheet-brass disk sample.

Reading	Continuous rotation of sample normal to shield X-Ray intensity (counts/sec)	Rotation only between readings X-Ray intensity (counts/sec)
1	8457	8343
2	8481	8498
3	8456	8410
4	8466	8357
5	8460	8454
6	8435	8504
7	8447	8380
8	8444	8489
9	8441	8386
10	8446	8468
Two standard deviations (95% confidence limits)	0.32%	1.45%

Table III. Comparison of the variation in x-ray intensity for molybdenum-coated cylindrical ceramic sample.

Reading	Continuous rotation of sample parallel to shield X-Ray intensity (counts/sec)	Rotation only between readings X-Ray intensity (counts/sec)
1	31341	31424
2	31352	31838
3	31370	34477
4	31292	33333
5	31352	31408
6	31356	31313
7	31377	34181
8	31376	34635
9	31390	32336
10	31369	31322
Two standard deviations (95% confidence limits)	0.17%	8.53%

Table II shows that for a fairly uniform sample, such as the pressed sheet-brass disk, continuous rotation improves over-all precision of the x-ray intensity measurement by a factor of four. The effect of a nonuniform primary x-ray beam and other geometrical effects are averaged by continuous rotation of the sample about an axis normal to the shield.

Table III shows that for a nonuniform sample, such as the molybdenum-coated ceramic cylinder, continuous rotation improves over-all precision of the x-ray intensity measurement by a factor of 50. A larger sample area of the nonhomogeneous sample is integrated by rotation about an axis parallel to the shield, thus providing the more representative measurement.

The advantages of the rotating sample holder are as follows:

(1) More reliable analytical intensity data can be obtained for nonhomogeneous samples.

(2) The sample holder is easy to operate, and it is easily assembled, disassembled, or adjusted.

(3) The sample holder can be designed to fit any sample characteristic. The only limitation is that sample and holder cannot be larger than the spectrometer drawer.

(4) The mechanism is compact since gearing is designed to fit into the holder with the sample and other fixtures.

(5) The contra-angle U may be oriented at any angle desired.

(6) The sample holder can be used without modification when samples do not require rotation.

Modification of a Vacuum X-Ray Spectro- 7.6
graph to Work in a Helium Atmosphere

W. De Spiegeleer and G. Vos

Analytical Chemistry Division, CCR Euratom Ispra, Italy

The use of a vacuum or helium atmosphere for the x-ray spectrographic analysis of long wavelengths is well known. Working with a vacuum path has the advantage of being a low cost technique but presents some drawbacks for obvious reasons, except when using special sample holders. For example, small quantities of powdered samples may scatter in the equipment when the air is admitted too violently into the chamber. For liquid or powdered samples, helium can be used, taking into account that it gives a lower intensity than vacuum for elements below sulphur in the periodic table.[1] However He is expensive, especially when it is necessary to have a continuous flow through the apparatus in order to wash away any trace of air. Hydrogen instead of He is another possibility and also cheaper, but much more dangerous.

A system without the above disadvantages was installed on the author's instrument (Philips Semi-Automatic X-ray Spectrometer PW 1220). It consisted simply of placing the equipment under vacuum, and introducing He gas slowly until a small overpressure of about 100-mm Hg was reached. To perform this operation, additional attachments were needed (see Fig. 1): an automatic safety solenoid valve (1) (Leybold cat. n° 27301, 10-mm diam), a diaphragm valve with direct way (2) (Leybold cat.

---

1. D. C. Miller and P. W. Zingaro, *Advances in X-Ray Analysis*, W. M. Mueller, Ed. (Plenum Press Inc., New York, 1960), Vol. 3, p. 49.

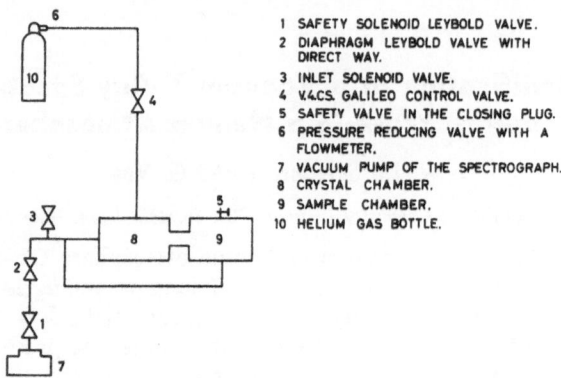

1 SAFETY SOLENOID LEYBOLD VALVE.
2 DIAPHRAGM LEYBOLD VALVE WITH DIRECT WAY.
3 INLET SOLENOID VALVE.
4 V.4.CS. GALILEO CONTROL VALVE.
5 SAFETY VALVE IN THE CLOSING PLUG.
6 PRESSURE REDUCING VALVE WITH A FLOWMETER.
7 VACUUM PUMP OF THE SPECTROGRAPH.
8 CRYSTAL CHAMBER.
9 SAMPLE CHAMBER.
10 HELIUM GAS BOTTLE.

FIG. 1. Scheme of the He attachment of the x-ray spectrograph.

$n°$ 17362, 10-mm diam), a control valve (4) (Galileo type V-4-CS), an adjustable safety valve (5), and a pressure reducing valve with a flow-meter (6). An Inlet valve (3) is included in the original apparatus and should be maintained closed, by hand, when the normal vacuum system is not used. A few simple modifications are required: replacement of the original "Tombac" piece connected to the vacuum pump (7) by valves (1) and (2) together with a rigid three-way tube, installation of an He inlet coming from gas bottle (10) through valve (6) and (4) in the cheek of the crystal chamber (8), and attachment of the safety valve (5) in the closing plug of the sample intake port. Valve (1) was electrically connected with the vacuum pump (7).

Procedure: (1) Load the apparatus with a pressed solid sample containing a high percentage of a light element (NaCl for example).

(2) Open valve (2) and close valve (4).

(3) When the apparatus is placed under vacuum, valve (1) is automatically closed to the entrance of air and allows the vacuum to pass.

(4) Apply the high voltage on the x-ray tube and set the goniometer at the 2θ value corresponding to the characteristic x-ray line of the chosen light element. Intensity of this line is recorded with the ratemeter.

(5) Open the He gas bottle (10) in order to obtain a pressure of about 1.3–1.4 atm, the flowmeter (6) being still closed.

(6) Close valve (2). (This is necessary, because when the vacuum pump is stopped, valve (1) is automatically closed, due to the low pressure in the apparatus. When He is introduced, the resulting small overpressure would open valve (1) and the gas would escape outside).

(7) Shut the pump down; valve (1) automatically closes the internal circuit.

(8) Slowly open the valve (4) and set the flowmeter at 3 liters/min to obtain a slight overflow of gas. A decrease of intensity of the x-ray line will be observed on the recorder.

(9) When the equipment is full of He, the intensity will return to its initial value. Set the flowmeter at 1 liter/min.

(10) After 2–3 min of stabilization, the spectrometer is ready for the analysis of liquids and light powders. It is possible to open the sample chamber to change samples without any loss of intensity, which is impossible when working under vacuum.

Operating time is roughly 10 min and a gas bottle of 10 m^3 under a pressure of 200 atm can be used 85 h. The gain of speed in comparison to working under vacuum with this equipment was about 30% when using counting times of 100 sec with three time counting repetition.

7.7   Simple Helium Enclosure for Increasing Diffraction-Line Detectability with a G.E. XRD-5 Diffractometer

Harold A. Johnson

X-Ray Crystallography Laboratory, UNIVAC,
St. Paul, Minnesota

Various means may be used to improve the detectability of an x-ray diffractometer, such as using pulse-height selection, etc. One method of increasing detectability is that of using a helium path instead of an air path for the x-rays. A graph comparing the detectability of diffraction-line intensities in a helium path to that in an air path for various wavelengths is given in Fig. 1. Chromium K alpha ($CrK\alpha$) radiation is used for greater dispersion of large d spacings such as are encountered in many organic compounds.

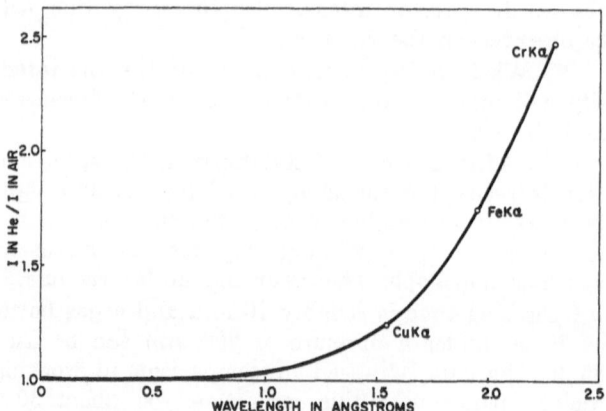

FIG. 1. Graph of ratios of helium-path to air-path diffraction line intensities vs x-ray wavelengths used.

The graph in Fig. 1 indicates that the detectability for CrKα is increased approximately 2½ times by using a helium path instead of an air path. A description follows for a simple method of constructing a helium enclosure for the G.E. XRD-5 diffractometer.

The first step in constructing the helium enclosure was to replace the G.E. diffractometer flat sample holder with another type. Figure 2 gives the machine-

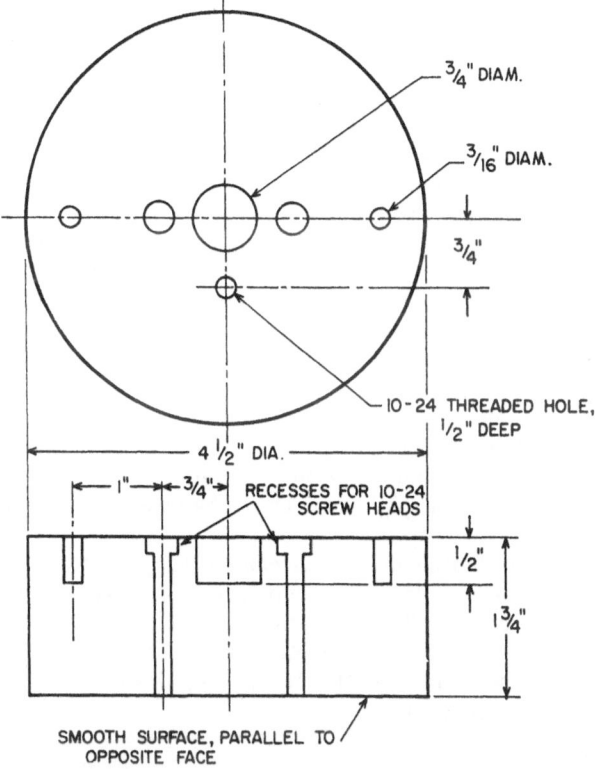

FIG. 2. Mechanical drawings of base for G.E. No. 2 SPG spectrometer x-ray fluorescence crystal holder used as diffractometer sample holder.

shop drawings needed for constructing the base of
the new flat sample holder from a $1\frac{3}{4}$-in. section of a
$4\frac{1}{2}$-in. diam aluminum rod. After mounting the base
in the center of the goniometer circle, a crystal holder
and radiation shield used for a No. 2 SPG x-ray fluorescence spectrometer and an SPG helium-tunnel
assembly (A 4961J) were obtained for mounting the
sample on the base and enclosing it in helium.

The second step was to make the two different
rectangular frames needed in order to attach the
helium-tunnel windows to the beam slit and soller
slit. Each frame was appropriately mounted in front
of its slit with two tiny screws, after drilling the
holes in the frames and tapping the holes for the
screws in the slit supports.

The third step was to wrap a small amount of
aluminum foil around the diffracted-beam slits extending from the helium-tunnel frame to the frame
holding the K_β filter. Cracks between the aluminum
foil and the supporting hardware were merely taped
shut with a good adhesive tape.

Thus, when the sample has been mounted in the
crystal holder and the air has been flushed out of
this system for about ten minutes with helium
through the nozzle in the back of the crystal holder,
the path from the x-ray tube window, to the sample,
to the K_β filter will have been changed from air to
helium.

It is a simple matter to change samples by turning off the x-rays and temporarily uncovering the
sample holder. The radiation shield with attached
rubber-helium tunnel assembly is removed by loosen-
ing the set screw in the front, and unsnapping the
windows from the frames.

Convenience Modification of the General 7.8 Electric X-Ray Spectrograph and Diffractometer

H. A. Johnson and M. L. Kinnaman

Materials and Process Engineering, UNIVAC,
St. Paul, Minnesota 55116

A simple method of modifying our General Electric x-ray apparatus and Leeds and Northrup chart recorder has been found to stop the chart recorder at the end of a goniometer scan even though the goniometer is left on unattended. After modification, the end of the goniometer scan corresponds to the end of the chart recording, thus avoiding confusion as to the two theta angles on the recording. Also, the chart paper is not wasted or unrolled unnecessarily when the operator is not present. The same type of modification was made on a Norelco x-ray diffractometer and spectrograph.[1]

Our instrument, obtained in 1961, is the Model XRD-5/F with a two-speed reversible goniometer motor. Also, an auxiliary chart-drive motor switch was provided near the top of the recorder. This G.E. x-ray unit, as purchased, had been changed at the factory so that figure 4 on addendum No. 1 to Direction 11 690C SPG spectrogoniometer is obsolete. For instance, there is a jumper from terminals six and seven and a capacitor between terminals one and five on the terminal strip below the goniometer dial panel. Before modification, the wiring under the goniometer dial panel was found to be as in Fig. 1.

The modification needed to stop the chart recorder from rotating when one of the limiting switches are

1. E. B. Buchanan, Jr., Norelco Reptr. **12**, No. 3, 98 (1965).

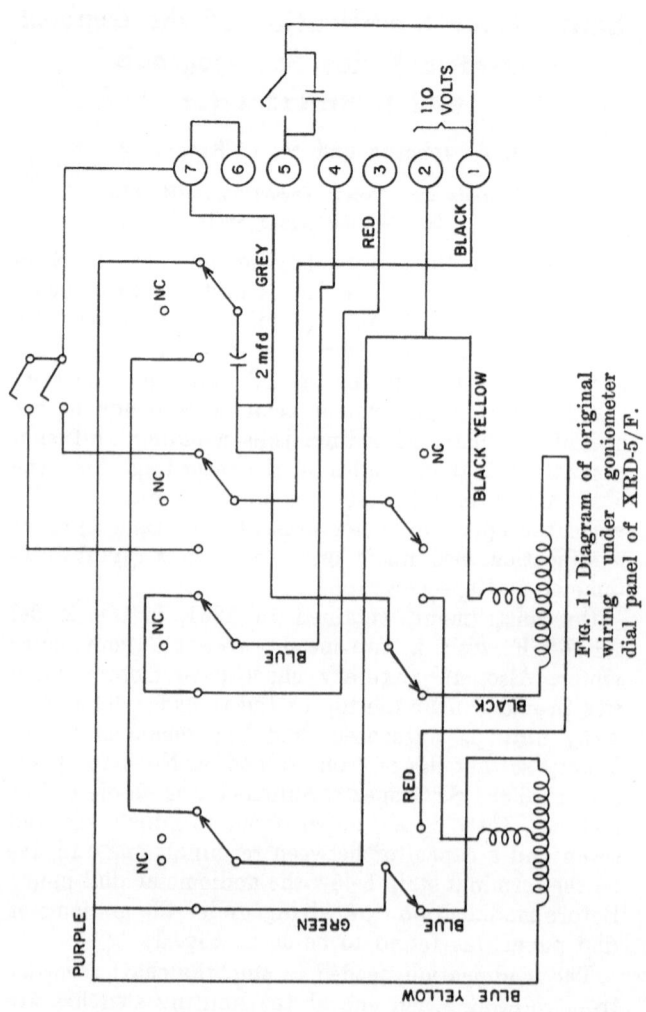

Fig. 1. Diagram of original wiring under goniometer dial panel of XRD-5/F.

depressed at the end of a scan consists only of a few simple rewiring operations under the goniometer dial panel and in the chart recorder. Anyone contemplating a similar modification should assure themselves that the wiring of their instrument is similar to that of our instrument.

The modifying procedure is as follows:

1. In the interests of safety, shut off the power supply to the instrument.

2. Lift off the top of the goniometer dial panel on the table, after unscrewing the number 10-32 socket-head screw in each end of the panel, and lay it upside down on the goniometer.

3. The red wire that comes from the recorder should be detached from terminal three and attached to terminal two of the goniometer terminal strip.

4. The blue wire that also comes from the recorder should be detached from terminal four and attached to terminal six of the goniometer terminal strip.

5. Replace the goniometer dial panel on the table.

6. Remove the two white wires from the lug on the terminal block above the chart-drive motor in the Leeds and Northrup recorder, solder them together, and tape. The white wire from the chart-drive motor should remain connected to the terminal lug.

7. A wire extends from the ''OFF'' side of the auxiliary chart-drive ON–OFF motor switch to terminal S1 on the terminal strip inside the recorder. Disconnect this wire from the S1 terminal and attach it to the above-mentioned lug on the terminal block above the chart-drive motor. Essentially, the modification consists of deriving the 110 V ac from the goniometer table connections instead of from direct connections in the recorder. A simplified diagram of the finished connections are given in Fig. 2.

8. When ready for use, turn on the power supply to the instrument, turn the auxiliary chart-drive

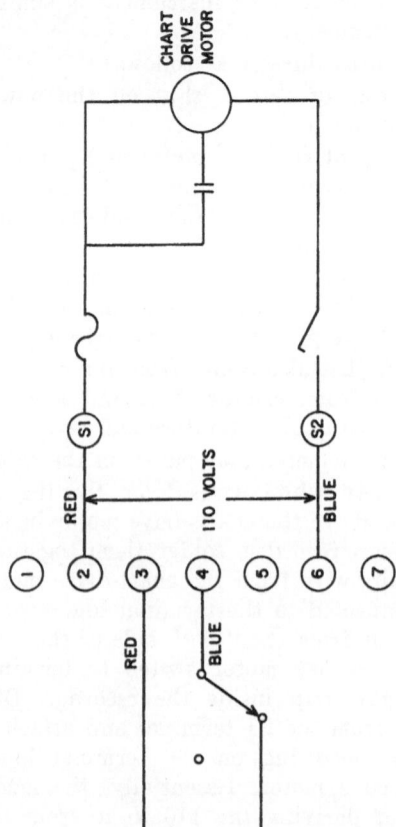

FIG. 2. Simplified diagram of wiring after modification.

motor switch to "on," and manipulate the increase–decrease goniometer switch to start a recording. To rotate the chart recorder alone, turn the goniometer motor switch to manual, and then manipulate the chart drive motor switch at the top of the recorder.

When the above operations are completed, a depression of the increase or decrease limiting microswitch on the x-ray spectrograph or diffractometer should stop both the goniometer-rotating motor and the chart-drive motor at the same instant instead of the goniometer-rotating motor alone.

Goniometer Stop for Use with Large Sample Chamber in G.E. X-RD5 X-Ray Spectrograph 7.9

Harold A. Johnson

X-Ray Crystallography Laboratory, UNIVAC,
St. Paul, Minnesota 55116

When the large sample chamber offered by Buick Motors is substituted for the regular sample holder with the General Electric X-RD5 x-ray spectrograph, the SPG preamplifier and counter chassis may strike the sample chamber when the goniometer is set for increase scan. With the regular sample holder and the stop bolt in the right-hand position, the goniometer stops at $148°$ $(=2\phi)$; in contrast, with the large chamber the counter chassis strikes it on an increase scan at $136°$. Substitution of the illustrated bumper (Fig. 1) machined from $\frac{1}{2}$-in. aluminum rod for the one provided with the goniometer increase-scan limiting switch stops the counter chassis at $136°$ just short of the large sample holder. The two existing 4-40

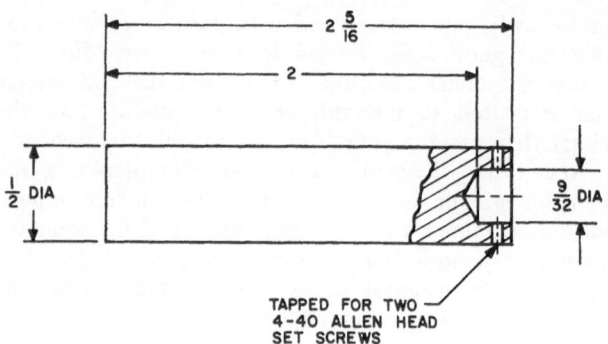

FIG. 1. Goniometer stop for use with large sample chamber.

Allen-head set screws need only be removed, the illustrated bumper substituted, the screws replaced, and the stop inserted in the usual 148° position.

SECTION 8

MISCELLANEOUS

Technique for Measuring the Emittance of Polymers at Elevated Temperatures[*] 8.1

J. G. Hansel

Guggenheim Laboratories, Princeton University, Princeton, New Jersey 08540

S. Y. Lee[†]

Combustion Laboratory, Stevens Institute of Technology, Hoboken, New Jersey 07030

A simple technique has been devised for determining the high-temperature, spectral emittance of materials which are melting and/or decomposing. The method has been applied to polystyrene; it could also be applied to some other condensed-phase materials.

Historically, emittance measurements were performed with materials which remain fairly stable at elevated temperatures. In some cases, unwanted chemical reactions and the corresponding surface changes could be prevented with the use of inert atmospheres, thereby permitting the surface character of the ma-

[*]This work is a portion of a larger research effort which was supported, in part, by the Office of Naval Research under contract at Stevens Institute of Technology.

[†]Present address: Westinghouse Electric Corporation, Research and Development Center, Pittsburgh, Pa. 15235.

SECTION 8

terial to be maintained during the measurements.
Stierwalt, and Riethof *et al.*[1,2] describe some of the
various apparatuses which follow this essentially
steady-state mode of operation.

In contrast to this, propulsion and re-entry researchers are often faced with the need for emittance
data of materials, such as polymers, which lose viscosity (or otherwise melt), decompose to lower-molec-
ular-weight fragments, or evaporate at the elevated
temperatures for which the emittance values are desired.[3,4] The various endothermic or exothermic reactions and the gradual disappearance of the sample
preclude any attempts to maintain the sample surface
at a constant temperature; thus, under these circum-
stances, the more conventional steady-state methods
of determining emittance are of limited usefulness.

The approach to be discussed here may be termed
a transient method, since no attempt is made to maintain the material surface temperature constant;
rather, the temperature is permitted to continually
increase and ultimately reach levels which would be
unattainable by steady-state techniques.

The following is a description of the apparatus and
the experimental technique. A small hot plate, fabricated from a coil of Nichrome wire and a circular
piece of asbestos, was used as a heat source. The
heating rate was regulated with a variable-voltage
transformer. The powdered polystyrene(in this case)

1. D. L. Stierwalt, AIAA Paper No. 65-654 (September 1965).
2. T. Rietholf, B. D. Acchione, and E. R. Branyan, ''High-
   Temperature Spectral Emissivity Studies on Some Refractory Metals and Carbides,'' in *Temperature etc.*, C. M. Herz-
   feld, Ed. (Reinhold Publishing Corporation, New York,
   1962), Part 2, p. 515.
3. J. G. Hansel and R. F. McAlevy, III, AIAA J. **4**, 841
   (1966).
4. P. L. Hanst, ''Surface Temperature Measurements on Ab-
   lating Missile and Satellite Heat Shield Materials,'' Ref.
   2, p. 489.

was placed in a shallow, thick-bottomed, brass dish, which was about 2.5 cm in diam and rested on the hot plate. A thermocouple was attached to a device which permitted the thermocouple junction to be raised or lowered near the surface of the material. The thermocouple leads were bent a few millimeters from the bead to facilitate the placing of the leads as well as the junction onto the surface and, thereby, reduce heat conduction loss into the leads. The procedure was to supply a steady current to the hot plate; and, after the polystyrene melted, the thermocouple was placed on the molten surface. With continued heating, the radiation intensity was recorded (Perkin–Elmer model 112 monochronometer) as a function of thermocouple emf until the last of the polystyrene decomposed to styrene, etc., which in turn vaporized. In some runs, temperatures of over 400°C were achieved before the sample disappeared (about 2 min). At temperatures above 250°C the rate of degradation of polystyrene is appreciable, and the viscosity becomes relatively low. These effects permit the styrene, etc., vapor which forms below the surface to escape readily, and, in doing so, create motion in the liquid polystyrene to minimize temperature gradients. In this respect, the exact location of the thermocouple in the pool relative to the surface becomes relatively unimportant.

In this type of experiment, it is essential that the molten material be relatively opaque at the wavelength being examined. If this is not the case, radiation from below the surface and the bottom of the brass dish will give a false value of the emittance. At the particular wavelength used for polystyrene (3.42 μ), the material is virtually opaque (absorption coefficient greater than 650 cm^{-1}). The emittance was determined to be 0.95 ± 0.05 at 400°C, which in this particular case is the same as the room temperature value. The blackbody source was a Barnes Engineering model 11-200.

8.2   A Cooling System for a Laboratory Magnet

R. A. Forman and T. McKneely

National Bureau of Standards, Washington, D. C.

In many fields of experimental physics, for example, in magnetic resonance, a large magnet is used which requires a clean, cool water supply for cooling purposes. It has been our experience that, during the summer months, the city water temperature often rises so high that the magnet cannot be operated at maximum fields for prolonged periods. Further, there is some question as to the advisability of running city water through the cooling coils of the magnet, as, even with filtering, some constriction of the coils with sediment was noted in our system.

With the above in mind, we designed a system capable of cooling a 5 kW magnet (in our case, a Varian V-4012A magnet) and using only distilled water in the cooling coils. This system has performed satisfactorily in our laboratory for a period of one year, requiring no maintenance other than the repair of a refrigerant leak in the water chiller.

Our application was to a magnet used for wide-line nuclear magnetic resonance, and this imposed several restrictions on the system. Of major importance was the fact that the system had to be capable of operating at different loads, inasmuch as the magnet is used at different fields. Equally important was the requirement that the cooling water reach an equilibrium temperature in a short time and that this temperature not fluctuate. Other attempts at producing cooling systems which had come to our attention did not satisfy the above criteria. These sys-

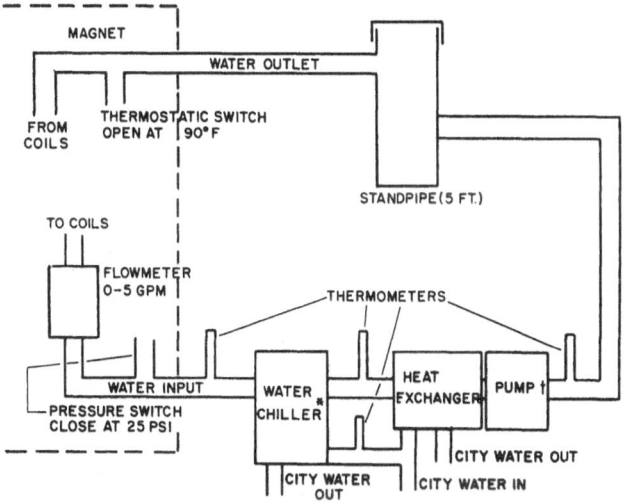

FIG. 1. Block diagram of cooling system. *Dunham–Bush Model CCP-40 w/capacity control †Pump Jacuzzi QS5RP/B (Spec. 3380, or equivalent)

tems were designed with a cooler which cycled on and off with demand. This produced electrical noise and also periodic fluctuations of the magnetic field.

A block diagram of the system appears in Fig. 1. The water chiller has the unusual feature of a by-pass in the refrigerant loop which effectively reduces the capacity of the chiller when the demand is low. This is important as the system is always electrically on, and no electrical transients are introduced as would occur if the refrigeration compressor were to cycle on and off. The additional external heat exchanger is used instead of the chiller when operating at low fields. A temperature interlock has been added to the magnet water system to turn off the magnet if the water temperature rises above 90°F at the output. This is in addition to the pressure interlock already a part of the magnet. A flowmeter has been incorporated in the system to monitor the water flow.

The over-all performance of the system has been quite satisfactory. The acoustic noise introduced into the room is approximately the same as a window air conditioner. The water temperature stability *at a given field* is quite high; the water temperature could not be observed to vary on the thermometers used, setting an upper limit on temperature fluctuations of 0.5°F. We feel the fluctuations of the system are much lower than that. It is to be noted that due to the nature of the water chiller, the equilibrium water temperatures will change with load. The change in temperature is approximately 6°F for each 25% change in the load. For application to high-resolution spectroscopy, we note that the equilibrium water inlet temperature to the magnet can easily be set over a wide range from 40° to 75°F at maximum load, but this is a job for a refrigeration engineer. The pump in this system was selected for its excellent noise characteristics.

The cost of the parts for the system was approximately $1000. Our total installed cost was approximately $1500. The system as installed required 20A/208V 1 φ and a maximum of 3 gpm of water at less than 85°F. The system occupies floor space of 2×4 ft with the major portion of the space used for the chiller.

The system as designed will handle the cooling for a 5 kW magnet requiring 2 gpm water flow. By use of larger chillers of the same type, large magnets could easily be accommodated. The basic advantages of the system as designed are: a closed distilled water loop for the magnet, freedom from temperature fluctuations (and hence field fluctuations) of the magnet, and the lack of production of electrical noise, particularly transient electrical noise.

Method of Cutting a Diffraction Grating 8.3

F. R. Lipsett

National Research Council of Canada, Ottawa 7, Ontario

Owing to an error in specifications, a plane-reflection grating recently received in our laboratory was found to be too long. This caused vignetting of a light beam in the monochromator in which the grating was to be used, with a consequent loss of transmitted light. Because of the cost and delivery time incurred in obtaining the grating, it was therefore desired to cut it. It was, of course, essential that the ruled surface of the grating remain undamaged during the cutting operation. Since a diamond saw was to be used, it was also essential that the lubricant used with the saw be prevented from reaching the rulings. The lubricant contained a wetting agent which penetrated most common adhesives, and finding a satisfactory adhesive was one of the main difficulties encountered. However, after a number of trials the following method of cutting the grating without damage was found.

The grating was a replica deposited on a glass blank $78 \times 58 \times 16$ mm. A plastic cover (Fig. 1) was made of Lucite in such a way that the sides, one end, and a 1-mm strip around the ruled surface of the grating came in contact with it. The grating was held in the plastic cover by masking tape on the sides. The portion of the ruled surface left exposed by the plastic cover was scraped off the blank with a

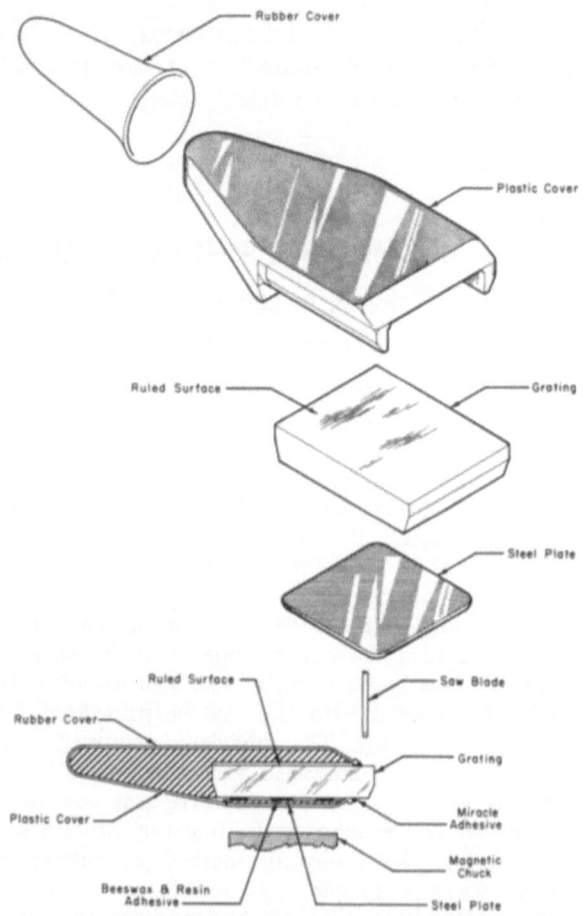

FIG. 1. Diagram of grating and protective assembly.

single-sided razor blade. A steel plate 1.6 mm thick
was then cemented to the bottom of the blank with
a beeswax and resin mixture of very low melting
point. The end of a rubber cover[1] was cleaned with

1. Cant Rip style. Novelty Rubber Company, Hamilton, Ontario.

ethanol and it was unrolled over the plastic cover, grating, and steel plate as indicated in Fig. 1. The rubber cover was a condom thick enough (0.20 mm) to withstand the handling necessary for cutting. The rubber cover was then cemented to the grating with two coats of Miracle Adhesive.[2] When the adhesive was dry the assembly was held on a magnetic chuck and the required length of the grating was cut off. After cutting, the covers were removed and the grating was ready for use. The portion of the rulings beneath the adhesive was of course no longer usable.

Two trial glass blanks of size identical with the grating, and the grating itself were cut. No difficulties were encountered, and no damage was done to the ruled surface.

Grateful acknowledgment is made to J. N. Cairns for helpful suggestions and for doing the cutting, and to J. Ferris[3] for helpful discussions.

Funnel Used for Powder Transfer 8.4

J. Trafford

Mullard Magnetic Components, Southport, Great Britain

At the start of this work the method of mixing samples with buffer and diluent for subsequent arcing in a carbon cup was to vibratory mill in a steel tube, transfer to a plastic tube, and pour, via a separate funnel, into the carbon cup electrode.

The sequence can be both shortened and simplified by vibratory mixing in a plastic tube without balls. A glass funnel is then attached to the plastic tube

2. Miracle Adhesive Corp., Bellmore, New York.
3. Bausch & Lomb Optical Company, Rochester, New York.

and the mix transferred directly to the carbon cup electrode. Variable bulk density in the mix is greatly reduced to give a smoother burn on arcing and because there are fewer operations involved, the risk of contamination is reduced.

If, for example, barium carbonate, carbon, and the sample (e.g., iron oxide) are to be mixed, the carbon is sieved through a 300-mesh nylon sieve. The barium carbonate and powder sample are then sieved through separate 100-mesh nylon sieves. Any sample remaining on the sieve is ground until it passes through the sieve. Weighed portions of the three constituents are transferred to a polystyrene tube (5.5 cm long × 1.2 cm diam) which is sealed with a polyethylene cap and clipped into an aluminium tube, which in turn is attached to a small mixer/mill.

The mixer/mill is not now commercially available but is similar in size to the Spex Mixer/Mill model 5000. It holds a 5.5-cm tube in the vertical position compared to a 4.8-cm tube in the horizontal position in the Spex Mixer/Mill.

The mix is vibrated for five minutes after which the cap is removed and a glass funnel attached to the tube (see Fig. 1). Coarse materials are previbrated with balls before sieving. The mix is then

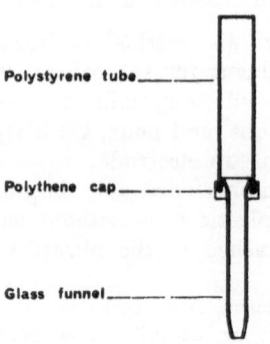

Polystyrene tube

Polythene cap

Glass funnel

FIG. 1. Tube with funnel attached.

Table I. Percent sodium carbonate in fine powder mixes.

Vibratory mixed	Ball milled
49.45	49.85
49.52	49.71
49.49	49.73
49.26	49.72
49.90	49.49
49.94	49.50
	49.61
Mean 49.59	49.65
1σ 0.27	0.11

poured via the funnel into the carbon cup electrode
and packed by gently tapping the electrode. More
mix is poured in until the cup is full. The side of
the funnel is used to level the powder and the mix
is arced. A funnel is permanently attached to each
tube of standard mix, but for the samples the funnel
is detached when the analysis is complete and cleaned
with a pipe cleaner ready for reuse. When duplicate
mixes are used, the same funnel suffices for both. The
end of the funnel is sealed with a rubber "police-
man" when not in use.

Mixtures of equal parts of finely ground carbon
and sodium carbonate were prepared by the old and
new methods. The homogeneity was checked by ti-
trating several weighed portions of each with hydrochloric acid to determine the sodium carbonate content. From the results in Table I the difference in
mixing efficiency was not sufficient to significantly
alter the analysis at the 0.001%–0.5% impurity
levels in which we are interested. Vibrating without
balls gave a uniform mix without the lumps which
characterized the old method. This in turn provided a steadier and more consistent arc and therefore, more accurate and reproducible analysis. Mixtures of equal parts of coarse alumina and sodium

Table II. Percent sodium carbonate in coarse powder mixes.

Vibratory mixed	Ball milled
45.26	52.30
43.18	47.58
38.65	45.77
28.06	52.29
38.74	52.38
61.10	53.69
53.01	46.37
72.31	38.35
Mean 47.54	48.59
1σ 14.08	5.16

carbonate were also prepared by both methods. The mixes were divided into a number of parts and titrated as before. From Table II the ball-milled result is observed to be better than the vibratory-mixed result but neither was satisfactory.

The mixing process works well with fine materials but coarse materials must be reduced to pass a 100-mesh sieve.

Shearing and Crushing Device for Mixing Powders*

8.5

Peggy L. Stewart

The University of Tennessee, Knoxville, Tennessee

Isabel H. Tipton

The University of Tennessee, Knoxville, Tennessee, and Health Physics Division, Oak Ridge National Laboratory, Oak Ridge, Tennessee

An inexpensive way to reduce granular samples of ash to a fine powder (75–150 μ particle size) has been developed by modifying a device suggested by Dutton.[1] A piece of 20-mil tungsten wire, formed into a spiral in the shape of a cocoon around a $\frac{3}{8}$-in. tungsten carbide ball (see Fig. 1), is placed with the sample in a plastic vial with a polyethylene cap and agitated on a mixer mill. The dimensions of the spiral are not critical. It should slide freely inside the vial and should be at least half as long as the vial. Since

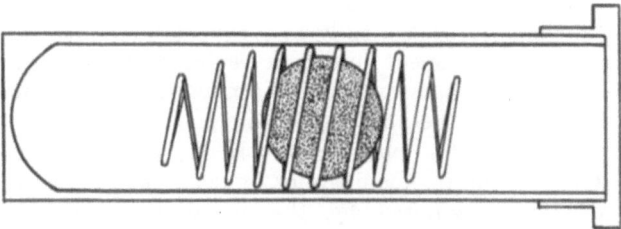

FIG. 1. Ball-in-spiral shearing device confined in plastic vial.

*This work was supported by the Atomic Energy Commission through subcontract 2351 under W7405 eng 26 between the University of Tennessee and Union Carbide Nuclear Corporation.

1. W. L. Dutton, Appl. Spectry. **15**, 24 (1961).

the plastic containers can withstand only about 20–30 sec of pounding at one time, agitation times should be kept to about 15 sec and the agitation repeated if the sample is not reduced to a powder in that length of time. After use, the ball-spiral device is removed with forceps, washed with a brush in dilute HCl, rinsed in distilled water, and dried and stored in an oven where it is kept ready to use again.

8.6 Simple Slit Servomechanism for the Rudolph Spectropolarimeter

David Moorcroft and Neal L. McNiven

The Worcester Foundation for Experimental Biology, Shrewsbury, Massachusetts 01545

The Rudolph spectropolarimeter[1] (model 656) in our laboratories is equipped with a dual monochromator containing entrance, intermediate, and exit slits.

In operation, as the wavelength was scanned continuously downwards, it was necessary to increase all three slitwidths manually at frequent intervals. This manual adjustment has been replaced by a simple slit servomechanism which permits the unattended operation of the spectropolarimeter.

Details of the modification—In the unmodified instrument the photomultiplier dynode voltage is automatically regulated by the signal level. This is achieved by the use of a Variac autotransformer, driven by servomotor A (Fig. 1), which is actuated by a signal derived from the photomultiplier by means of a chopper amplifier. The existing circuit

1. Rudolph Instruments Engineering Co., P. O. Box 265, Little Falls, N. J.

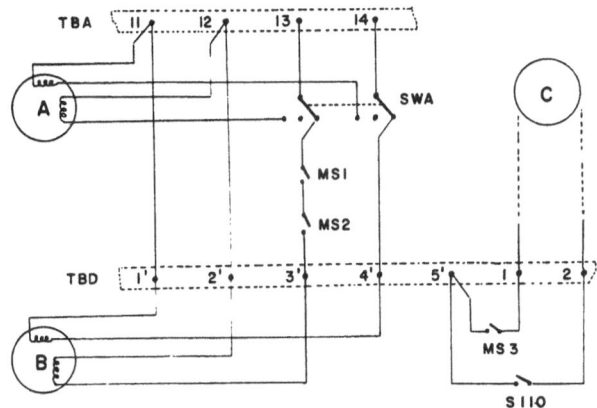

FIG. 1. Modified electrical circuit.

was modified to direct the output of this amplifier via switch SWA (double pole, double throw, center off) to servomotor B (Minneapolis Honeywell type M623 AY3 × 1) of a similar impedance to servomotor A. Servomotor B is used to vary the slitwidths. The switch SWA permits the original photomultiplier voltage regulating system to be used if required.

The knurled hand wheels controlling the monochromator entrance and exit slits were removed and replaced by two sprockets (Boston Gear Co. No. GA 20). Details of the method of replacement are given in Appendix A.

These sprockets are simultaneously chain driven by servomotor B, mounted in front of the monochromator. A piece of adhesive-backed Teflon tape was inserted in the trough on the top of the monochromator to reduce chain friction.

Limit microswitches MS1, MS2 set for slitwidths of 0.02 and 1.98 mm, respectively, were installed. It was also found convenient to install another microswitch MS3 in the wavelength drive circuit to inter-

rupt the scan motor C when the slitwidth reached
1.90 mm. This microswitch is in series with the existing lower wavelength limit switch S110. Microswitches MS1, 2, and 3 are activated by the movement
of the exit slit actuating arm. This was achieved by
removing the existing retaining set screws of the slit
arm and replacing them with fillister-headed screws.
The microswitches were mounted on a bracket on the
front of the monochromator, inside the outer casing.
Terminal board TBA was already in the instrument
electronic cabinet, and TBD is a 9 position replace-

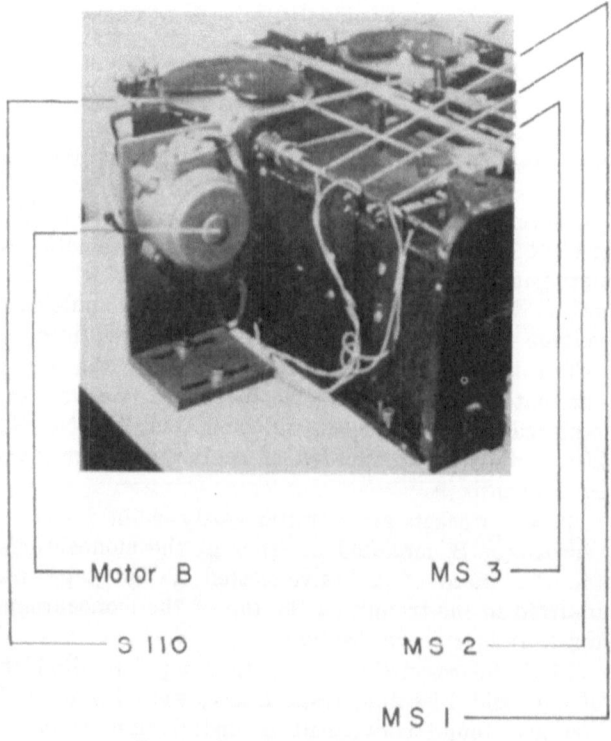

Motor B   MS 3

S 110   MS 2

                                    MS 1

FIG. 2. Completed slit servosystem.

ment of the original located on the left-hand side of the monochromator.

The completed assembly is shown in the photograph in Fig. 2 and modifications to the electrical circuit in Fig. 1.

Operation—Operation at a fixed photomultiplier voltage, using the servo-slit is achieved by opening the intermediate slit to 2 mm and switching in the photomultiplier servomotor A with SWA. By adjusting the entrance and exit slits manually, the desired photomultiplier voltage is obtained on the panel meter. Servomotor B is then switched in by SWA. It was found that leaving the intermediate slit set throughout the run at 2 mm rather than progressively adjusting it produced little or no effect on the spectrum.

Results—The ORD spectra of (a) sucrose and (b) 5α-Androstan-3,17-dione were determined under the following conditions to evaluate the new system.

Method 1: Servomotor A activated (photomultiplier voltage being now regulated). All three slitwidths were adjusted stepwise by hand, as the following wavelengths were reached during the scan: 450 mμ, 0.23 mm; 400 mμ, 0.31 mm; 350 mμ, 0.50 mm; 300 mπ, 0.83 mm; 275 mμ, 1.08 mm; and 250 mμ, 1.67 mm. This gave a maximum half-intensity bandwidth of 5 mμ (see instrument operating instructions).

Method 2: The photomultiplier panel meter voltage was set to 75 V with the intermediate slit wide open at 2 mm throughout the scan. Servomotor B, controlling the entrance and exit slits was activated. The slitwidths were below the values given above and therefore the half-intensity bandwidth was less than 5 mμ.

(a) Sucrose: 26.5 mg/ml in water in a 1 cm long, 1-cm i.d. tube. Scan speed 5 mμ/min. Five results

Table I. Rotation of sucrose solution.

Wavelength mμ	Method 1		Method 2	
	Rotation (degrees)	Confidence[a] limits P, (0.99) n, 5 ±	Rotation (degrees)	Confidence[a] limits P, (0.99) n, 5 ±
450	0.329	0.007	0.332	0.007
400	0.424	0.016	0.422	0.011
350	0.568	0.014	0.572	0.012
300	0.831	0.016	0.828	0.017
275	1.045	0.018	1.036	0.020
250	1.392	0.021	1.375	0.022

[a] ASTM Manual on Quality Control of Materials, Publication 15C (January 1951), p. 50.

were taken alternately by each method and are shown in Table I.

(b) 5α-Androstan-3,17-dione: 2.0 mg/ml in dioxane in a 1 cm long, 1-cm-i.d. tube. Scan speed 5 mμ/min. The results obtained by the two methods are shown in Fig. 3.

These results show that, in the region 450 to 250 mμ, there is no statistically significant difference in the values, or the precision, of rotations measured by the use of this new device. This modification probably could be applied to other models of Rudolph spectropolarimeters.

This investigation was supported by Public Health Service Research Grant No. 5-S01-Fr-05528.

Appendix—Details of the method of replacement for each of the hand wheels.

(a) Set slit to or near the closed position, and loosen the screw in the hand wheel.

(b) Remove slit counter assembly.

(c) At the right-hand side of slit assembly, the slit movement arm runs through a guide block. Remove

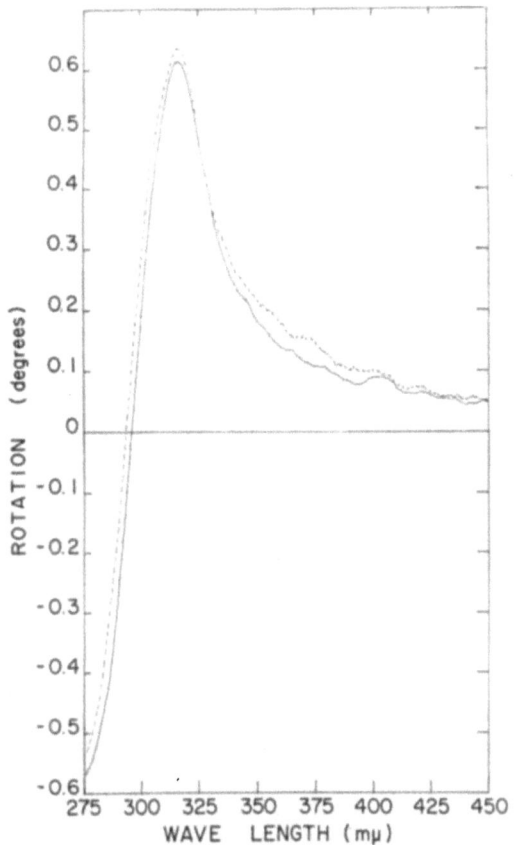

FIG. 3. ORD spectrum of 5α-Androstan-13,17-dione; ——
method 1, ---- method 2.

the tensioning spring and the two screws securing the
guide block to the monochromator case.

(d) Remove the screws from the bearing block
located to the right of the hand wheel.

(e) Lift the slit arm assembly slightly to permit
removal of the hand wheel and replace with the
sprocket.

(f) Reassemble slit assembly in the reverse order.

APPLIED SPECTROSCOPY
REFERENCE INDEX

AUTHOR INDEX

CUMULATIVE SUBJECT INDEX

(Numbers in parentheses indicate volume in series)